# FARM
# TRACTORS

# FARM TRACTORS

APRIL HALBERSTADT • PHOTOGRAPHY BY HANS HALBERSTADT

**Metro**Books

# MetroBooks

An Imprint of The Michael Friedman Publishing Group, Inc.

Library of Congress Cataloging-in-Publication Data available upon request.

ISBN 1-56799-560-8


Editors: Stephen Slaybaugh, Celeste Sollod, and Reka Simonsen
Art Director: Kevin Ullrich
Designer: Diego Vainesman
Photography Editor: Deidra Gorgos
Production Manager: Karen Matsu Greenberg


Color separations by HBM Print Ltd.
Printed in China by Leefung-Asco Printers Ltd

10 9 8 7 6 5 4 3 2

For bulk purchases and special sales, please contact:
Michael Friedman Publishing Group
Attention: Sales Department
230 Fifth Avenue
New York, NY 10001
212/685-6610 FAX 212/685-1307


Visit our website:
www.metrobooks.com

# Acknowledgements

This book owes its being and its vitality to the hundreds of collectors who are carefully maintaining an important part of our heritage. The tractor collections of hundreds of enthusiasts are preserving American agricultural history better than ordinary museums.

We are especially grateful to the advisors, collectors, enthusiasts and historians who each put their two cents' worth into this project: Carmin Adams, Bill Bechtold, Jim Becker, John S. Black, John Boehm, Mark Bookout, Phill Christiansen, Glen Christoffersen, JR Gyger, Dick Harold, Dale Hartley, Donald Henderson, Charlie Hill, Everett Jensen, Dana Johnson, Terry Kubicek, Richard Kuckenbecker and the Kuckenbecker Tractor Company, Blake Malkamaki, Ira Matheny, Doug Peltzer, Augie Scoto, Bob Seith, Ben Silva, Dave Soper, Robbie Soults, Dean Vinson, Denis Van de Maele, and Hank Wessel.

This book is also a tribute to the efforts and contributions of those tolerant individuals who took time out of their busy schedules to roll out their glorious tractors and find just the right settings for the photography.

We also appreciate the continuing feedback of researcher Guy Fay and the ongoing support of Pat Ertel, editor of *Antique Power Magazine*.

Thanks to all. We hope you like the book.

—Hans and April Halberstadt
January 1998

# CONTENTS

# Introduction

Tractors are fascinating, both as machines and because of the history they represent. From the moment the first gasoline tractor appeared in 1901, every new model chugging across a field in agricultural North America represented some interesting technological improvement. Of course, there were a number of early tractors that were impractical or unsafe, but even these machines frequently had innovations that an enterprising engineer could analyze, refine, and improve.

Advancements in engines, transmissions, steering, and ignition systems all made the farm tractor a more useful tool. The development of such innovations as the power take-off system, the hydraulic lift, and the three-point hitch were tremendous advantages to the average farmer. Since better technology improved agricultural productivity, each advancement in turn had an impact on the American economy. Mechanical improvements came quickly, and each change influenced others.

But, history of technology aside, old tractors are interesting just to see and hear and smell. Many tractor enthusiasts have a certain "brand loyalty," preferring one tractor maker over another. Others are intrigued by tractors built before 1935, the year the first "styled" tractor appeared. And some enjoy rare and exotic tractors—orchard tractors, for example, or limited production tractors, machines built in very small numbers for a very limited market.

Putting aside personal preferences in tractors, consider for a moment just how tractors fit into the great cosmology of American agriculture. Development in the tractor industry depended on related businesses. We look only at tractors in this book, but it's important to keep in mind some of the other factors that influenced tractor development.

The gas-powered tractor was very dependent on improvements in the petroleum industry. Although the potential power of gasoline as a fuel was known in the nineteenth century, the quality and quantities of this critical fuel were not reliable. There were dozens of petroleum-powered engines patented before 1900, but few were put into mass production, simply because there was no adequate system to process and distribute this marvelous fuel. So before the average American farmer could benefit from this new invention, the tractor, entrepreneurs like John D. Rockefeller had to build a reliable gasoline distribution system for automobiles.

A good gasoline sales network was only one part of the equation. The fuel itself needed to be processed into standardized grades. We will talk about the "gasoline" tractor in this book, but to be quite correct, the petroleum product burned was what the old-timers called distillate; it is closer to kerosene than the fuel we now call gasoline. Tractors did not burn regular seventy octane gasoline, the same grade used for automobiles, until 1935. To be strictly accurate we should probably call the early machines "petroleum-fuel tractors," but the term gasoline tractor is common and close enough.

The early mechanics and inventors who built cars, trucks, and tractors had to understand a great deal about the properties and limitations of various grades of petroleum fuels. It should not come as a surprise to discover that Charles Hart, the co-developer of the first production tractor, was a petroleum engineer. The gasoline-powered tractor was a marvelous mechanical achievement; but so was the development of the petroleum industry that helped make it possible.

There were other factors that contributed to the early success of the gasoline tractor, and we should mention a couple of them. These factors are frequently overlooked because they cannot be seen or held in your hand. They are business networks, and they include the agricultural machinery distribution network, the advertising industry, and the credit industry. All three industries were already present when the gasoline tractor appeared, and all were to play a critical role in the survival and success of the tractor industry.

Perhaps the oldest and most important of the three was the agricultural machinery distribution network, sometimes referred to as the branch-house system. Each traditional machinery manufacturer, such as the J.I. Case Threshing Machine Company, which started building threshers in 1842, or the John Deere Company, which traces its history to 1837, had a nationwide network to sell farm equipment. This network was composed of regional sales centers and warehouses known as branch houses. Local sales representatives were in smaller towns throughout each branch-house territory.

An important adjunct of the agricultural machinery business was a credit system that allowed the farmer to pledge part of a future crop against the purchase of a new thresher or steam engine. Some of the big manufacturers had their own financing systems and offered the farmer a loan with the credit system "in-house," while others partnered with a local bank. The agricultural credit industry would help farmers buy tractors.

The branch-house distribution system also produced a network of local sales representatives who were field experts. The same fellows who sold the farm machinery were frequently in the field with the farmer, demonstrating their products and training the farmer to operate newly purchased equipment. One of the first and most famous salesmen was Jerome Increase Case himself, developer of the thresher. As early as 1850, Case could be found in the field demonstrating the operation of his machinery, calling on farmers who owed him money for machines they had purchased, and helping with mechanical difficulties when problems arose. Case's business practices became an industry standard.

By the turn of the century, all the big agricultural machinery conglomerates had mechanical experts within their ranks.

A wheat harvest on Frank Dyke's place near Thornton in central Washington State, August of 1929. This is the technology that tractors replaced, and in the first decade of the century, nearly all farmers harvested this way. Tractors gradually replaced the horse over a period of fifty years, for both economic and emotional reasons, although there are still die-hards farming with horses today.

Stationary "belt" work was just
as important to the farmer as a
tractor's ability to plow or pull
a binder or mower. Threshing
time was the high point of
the farmer's year—payday
for all those months of hope
and hustle.

These men were sales representatives with expertise in both farming and mechanics. Every major agricultural machinery house had troubleshooters who could quickly hop a train and repair machines in the field. These were the first "field engineers," and the kinds of services they provided are now acknowledged as part of the American way to do business. Companies now service the machines they sell: the idea started in the agricultural machinery business.

Advertising was also a mature business by 1900 and would play an important role in the success of the tractor industry. There were hundreds of agricultural newspapers and magazines in distribution in 1900, a surprising fact to contemporary Americans who can now recall only two or three. But America's oldest continuously published magazines were farm journals, and they had carried advertising for farm machinery for decades. It is reported that by 1900, some advertising agencies had already begun to specialize. At the turn of the century there was already at least one agency in Chicago, the nation's largest agricultural hub, that specialized in agricultural industry accounts. In addition, the telephone, the telegraph, and an enormous railway system provided a network of information for all agricultural businesses at the dawn of the twentieth century.

Petroleum industry advances, advertising and marketing, sales and service, consumer credit—these all helped the tractor industry thrive. So while the focus of this book is on the development of the tractor and its technology, it also touches upon the development of other industries.

The tractors in this book are the machines that have stood the test of time. Here to represent recognized engineering advances are many of the marvelous machines that made an important contribution to tractor history. Tractors over fifty years of age are considered antique, but most of these machines still operate and some can even do a full day's work. Some folks guess that there were two hundred tractor builders by 1920. Those builders are gone and the machines they built, tractors like the Four Drive, the Hollis, and the Square Turn, are nearly forgotten. Yet without them, there would not be the modern machines farmers use today.

Gathered here are the tractors not forgotten. They represent some of the most interesting and finest technology ever designed. These machines are the reason that less than two percent of the American population is now able to harvest the food for all the rest—and large portions of the world as well.

**Early gasoline tractors looked a great deal like the steamers they replaced. The heavy steel wheels and chain-operated steering system were nearly identical to old steam engines, but the boiler was replaced with something brand new. This wonderful OilPull is owned and operated by collector Irv Baker.**

# Chapter 1

# Sowing the Seeds— The Birth of Tractor Technology

## *(1900 to 1912)*

Before there were gasoline-powered tractors, there were steam engines and steam-powered traction machines. Steam was the power source of the nineteenth century. Big steam boilers fired with coal or wood provided the power to drive ships, railroad locomotives, industrial power plants, and farm machinery. Many thought the age of steam would never end, and even the venerable J.I. Case Threshing Machine Company was still manufacturing steam-powered threshers until 1924. There were thousands of steam traction machines in agricultural North America, widely used because their fuel, wood and coal, was readily available and easier to handle than petroleum products.

Farmers who ran small farms or who were poor, however, still used animal power rather than steam engines. Horses, mules, and oxen helped plow, plant, and harvest. Draft animals remained on North American farms until well into the twentieth century, before finally being displaced in the late 1940s by gasoline tractors.

## Gasoline Engines

Most history books credit Nikolaus A. Otto of Germany with the development of the gasoline engine. He built his first gas engine in 1861 and won a gold medal at the Paris Exposition in 1867 when he exhibited it to the public. He then made some important refinements and introduced the four-cycle engine to the world in 1876. The potential of this power source was immediately recognized by many interested parties. The Encyclopedia Britannica reports that thirty thousand Otto-type engines were built within the next ten years.

Retrospect often makes it easy to recognize the connections between seemingly random events, as well as making it clear that great innovations are rarely made alone—inventions and advances are made by legions of people, sometimes cooperating and sometimes competing. The eventual development of the gasoline tractor depended on the discoveries of several people.

Although Otto is credited with building the first workable gasoline engine, his patents were immediately challenged. In 1886, the Otto patent expired; a French engineer named Alphonse-Eugène Beau de Rochas was given patent credit. Nevertheless, since we know that Otto was the first to actually build a successful engine based on these principles, historians still refer to it as the "Otto engine."

When the Otto patent was overturned in 1886, dozens of manufacturers were eagerly waiting. They had already been trying gasoline motors as a power source in every imaginable application. Now they would be free to put gasoline-powered machinery into production without having to pay a royalty to Otto. The first gasoline engines were used to replace steam power, as stationary engines for pumps and similar contraptions. Small experimental gasoline engines were also bolted to carts and carriages, sometimes with disastrous results but sometimes with reasonable success.

The Waterloo Gasoline Traction Engine Company in Waterloo, Iowa, was just one of many small companies trying to match a gas engine to a suitable chassis. According to historian Randy Leffingwell, John Froelich of Waterloo reportedly tried mounting a Van Duzen gas engine to a Robinson tractor chassis in 1892, thus creating the Waterloo Boy Tractor. It was not a successful pairing. Although Froelich tried to build a few more gasoline traction engines, none of his modifications proved to be practical or reliable. Froelich sold his company and left the business by 1897.

Many historians believe that the Waterloo Boy Tractor should be considered the first real gas-powered traction engine, citing the fact that the early Froelich machine demonstrated its potential when it was used to power a thresher one summer. Others assert that it made no lasting impact

and, therefore, discount its importance. Nevertheless, it is important to consider some related events that were taking place in other Midwestern workshops before making a judgment about the Waterloo Boy's claim to fame.

In 1896, the little contraption that would come to be known as the Ford automobile was being assembled in a garage in Detroit. Henry Ford was not a lonely pioneer, however. Allan Nevins reports in his history, *Ford: The Man, The Times, The Company,* that in the year 1900 alone, thirty-eight companies were formed in the United States to build automobiles. Europeans were also active auto manufacturers, producing vehicles like the Daimler, the Vauxhall, the Panhard, and the Renault before 1900.

Many manufacturers, European and American alike, had seen the potential of the Otto engine. Interest in this new technology for farm machinery was not motivated, however, by its potential power or its economy. Steam engines already provided all the power most farmers needed for operating a threshing machine or feed mill. Steam-powered machinery was at the peak of its performance near the end of the nineteenth century, and most small farmers were quite content to continue using horses for plowing and cultivating.

The motivation for the overwhelming interest in the gasoline engine was efficiency and, of course, the potential riches that could come to the successful inventor of a popular and profitable gasoline machine of any sort.

Compared to a steam engine that can produce the same amount of horsepower, a gasoline engine is very small and compact and weighs much less. Lighter weight was an important consideration when ponderous steam tractors were routinely crashing through flimsy rural bridges and crushing farmers and tractor operators. Gasoline engines were also less likely than steam boilers to explode.

In addition, the gasoline engine offered a simplicity and economy of operation that was not possible with either horse or steam power. A steam engine uses tremendous quantities of both fuel and water, and typically needs two or more crewmen to attend to it while in operation. Horses also need tremendous amounts of attention, as well as frequent rest periods, and consume fuel every day, whether they are working or not.

A gasoline engine has several advantages: it only needs a full fuel tank and some critical parts oiled; it takes only one operator; and it can be put in the barn and left for weeks without requiring attention. Gas engines were a clear improvement over other power sources.

More than anything, though, the gasoline engine was something new and exciting. Internal combustion was a novelty; steam power had been around for decades. Let us not forget that the history of civilization is chock full of moments like this, when the new is preferable to the old simply because of the novelty.

Charles Hart and Charles Parr were just two of the bright-eyed engineers who saw some of the new possibilities of the gas engine. They had read all of Otto's patent documents and were eager to develop their own applications for this new technology. When John Froelich became discouraged, selling his Waterloo, Iowa, tractor company and moving on, some miles north, in Madison, Wisconsin, Hart and Parr were still working on their new machine.

# 1901: The First Tractors— Hart-Parr

The credit for developing the first gasoline tractor in America goes to a pair of young inventors working in a university shop. It wasn't called a "tractor" yet; they called their machine a traction engine. But within a decade, it was clear to everyone in the agricultural machinery business that Charles Hart and Charles Parr deserved the title of Founders of the Tractor Industry.

Hart and Parr, inventors of the first production gasoline-powered tractor, were college students at the University of Wisconsin in Madison in the 1890s. Like many college students, they were visionaries. Intense, enthusiastic, and fueled by imagination, they managed to spread their enthusiasm for gasoline engines to their professors and their classmates.

While still attending agricultural college, the two designed, developed, and built an extremely reliable stationary engine. This motor proved so remarkable that Hart and Parr set up a company in 1896, before they graduated, to manufacture and sell their engines.

A number of university instructors and friends provided the necessary capital to build the first Hart-Parr factory in Madison. Their first product, an oil-cooled stationary engine, was sold to the *Fennimore Times* in Fennimore, Wisconsin, to provide power for the newspaper plant. It remained in continuous operation for more than twenty years. With such demonstrated reliability, Hart and Parr had no trouble finding customers for their machines.

But the young inventors were soon at odds with their more mature investors. The pair wanted to build portable engines. The investors, looking at the obvious success and outstanding profitability of their present production line, wanted to continue building the large stationary power plants. Charles Hart discussed the dilemma with his father during a visit home. The elder Hart suggested that Charles talk with a local banker and family friend, Charles Ellis. Ellis quickly arranged a loan of $50,000 for a new venture.

Hart and Parr moved their business in 1899, setting up a new corporation and building a new factory in Hart's hometown of Charles City, Iowa, in 1900. They quickly built and tested a prototype tractor for a local buyer. The first Hart-Parr tractor was built, tested, sold, and delivered to David Jennings of Clear Lake, Iowa, by the middle of 1901.

The second tractor, also a prototype, had a Hart-Parr engine mounted on a slightly different frame. In addition to a modified frame, a planetary clutch, (an early drive-train assemblage), was introduced on this machine. The motors for both machines were thirty horsepower engines. This tractor also had an immediate buyer, a South Dakota farmer.

With these two successes, the new Hart-Parr Company was ready to put its machines into limited production. In 1902, they built the first of a series of thirteen tractors. The first advertisement for a Hart-Parr traction engine appeared in the *American Thresherman* news magazine in 1902.

History was being made in the small Iowa town. Not only were Hart and Parr producing the first modern traction engines, they were building them on what would later be called an assembly line, making the Hart-Parr Company one of the earliest manufacturers to use production engineering and assembly.

Manufacturing and milling a number of sub-assemblies at the same time, and then building several machines at once, was an historic milestone. Hart and Parr were using an entirely new method of producing machinery, procedures they

ABOVE: Every mechanic with a good idea built a tractor, hoping to cash in on the demand for a good, inexpensive machine. This offering is a "tricycle" with front wheel drive, an early innovation. The bench seat is an interesting feature, too. But it would take a very tall driver to see the field in front of this tractor.

**Charles Hart and Charles Parr established their tractor factory in Charles City but the name was just a coincidence. This northern Iowa town can proudly claim to be the birthplace of the American farm tractor.**

had used in their earlier plant in Madison. Setting up the lathes and milling machines to produce several tractors at once allowed economies of time and labor. Although the theory of assembly-line production was known, the Hart-Parr Company is considered to have pioneered one of the earliest practical demonstrations of this important innovation.

Tractor sales started unevenly, then began to climb. Hart-Parr sold 37 tractors in 1903 and 23 in 1904. The company did better the next year, selling 51. Then, in 1906, sales escalated to 170. Hart-Parr was building a reliable machine, but a great deal of its sales success was due to the company's use of advertising in journals like the *American Thresherman*.

W.H. Williams, a classmate of Hart and Parr at the University of Wisconsin, had joined the com-

pany as sales manager in 1905. It was Williams who popularized the word "tractor" by using it in advertisements he wrote for the company. And it was Williams who first named and marketed the concept of "power farming," a system of efficiently organizing farmwork by using a gasoline tractor and matching implements.

Power farming started as an advertising slogan and a marketing campaign but quickly became a widespread concept. By 1910, nearly every agricultural machinery builder had picked up the phrase and incorporated it into their own advertising. This concept of mechanized farming introduced remarkable new efficiencies by letting the farmers control more of their production. Even the popular agricultural magazines of the time began writing articles about this new idea: planning farm production around the use of the tractor.

With a versatile Hart-Parr tractor and a set of implements, a farmer could quickly gather a crop if he saw a storm coming. It was also now possible to plow or cultivate without having to stop and rest or water the horses. Before the Hart-Parr engine, agricultural power had been provided by steam engines. Dangerous and heavy, a steam behemoth required a trained operator, a large load of fuel, a good supply of water, and a considerable number of crewmen to keep it moving.

In contrast, the Hart-Parr tractor with its gasoline engine could be managed by a single operator. Here was a machine that most farmers could operate by themselves, stopping whenever they felt like it.

The Hart-Parr Company would have the tractor market to itself for nearly a decade. By about 1908, however, serious competition started to arise. The International Harvester Company was developing a type of tractor they called a "cultivator." It was powered with "friction drive," an early type of transmission technology that would prove useful for more than a generation. Henry

Ford began demonstrating some tractor prototypes and a motor cultivator of his own by 1906. Department of Agriculture statistics estimate that there were about six hundred gasoline tractors in America by 1910. More than half of them were built by Hart-Parr, but the remainder were built by more than two dozen other makers who had ambitions to be tractor producers.

## Steamers Meet Tractors

By 1908 there was a lot of interest in gasoline-powered traction engines, but there was no clear idea what one should look like. In fact, even the terminology was just beginning to sort itself out. The word "tractor" first began appearing in 1906, courtesy of the Hart-Parr Company. Until that time, the invention was referred to as a "gasoline-powered traction engine," a real mouthful. This term was derived from the name "steam traction engine," the machine that most folks knew well.

The most common type of powered machinery on the farm before 1900 was the steam engine,

also called a steam traction machine, a fire-breathing, smoke-belching boiler the size and shape of a small locomotive. Steam engines came in two varieties: one that could move under its own power, called a traction machine, and one that could not, referred to as a stationary engine. The word "traction" means to drag or to pull.

Steam engines for farm work had been developed around 1876. The earliest models could not operate under their own power, even though they were on wheels. These heavy boilers were dragged to the field by teams of draft horses and fired up on location. They were then belted to a thresher, providing the power needed to harvest a wheat crop.

It was a tremendous advance in technology in 1884 when these giant steamers were fitted with a drive shaft, actually allowing them to move from farm to farm under their own power. There were thousands of steam engines built for use on farms, and they would continue

to be manufactured until the 1920s. The J.I. Case Threshing Machine Company produced 36,000 steam engines, two-thirds of the total number built in the first two decades of the century, finally ceasing production of the behemoths with its last run of 134 steam traction engines in 1924.

Steam traction machines were quickly adapted to pull all types of heavy loads. They were used for logging, for pulling large plows and breaking heavy prairie sods, and for towing tons of construction material.

Steamers moved very slowly, two or three miles per hour (3.2 or 4.8kph), but they did have tremendous power. For example, they could easily pull five or six gang plows, which each weighed approximately two tons (1.8t), at one time. Or they could be belted to a threshing machine, providing the power to operate the thresher. Enormously expensive, these large steam traction machines were sold primarily to grain farmers of the northern Midwest states who had thousands of prairie acres in production.

So when most inventors first thought about developing a new gasoline-powered traction machine, they thought of modifying a steam machine, fitting a gasoline engine to a steamer chassis. It was a logical progression. Looking at many of the early tractors, we see that they resemble small steam engines.

Steam machines would also leave their imprint on the model designations of tractors for at least two decades. Steam tractors were sized or rated by two horsepower designations: the amount of horsepower produced when the machine was pulling a load and the amount of horsepower available when standing still, powering a thresher. This rating system became somewhat formalized at the Winnipeg Industrial Exhibition in Canada in 1909. So a Model 25-45 would be the designation of a machine that tested at twenty-five horsepower pulling a plow and forty-five horsepower powering a sawmill or thresher. And, therefore, a Model 12-25 would be about half the size of the more powerful 25-45. It became a meaningful way to compare the output of steam machines and gas machines alike.

But during the first decade of the century, the terminology and the technology were still being sorted out. Gasoline-powered machines would soon become known as "tractors"; the older, steam-powered machines would be commonly referred to as "steamers." Model rating numbers would disappear completely by the 1930s, replaced by catchy names like "Farmall" or "Do-All." Tractors would become small and agile, capable of performing several farm operations simultaneously. Still later they would become quite large again, rated at 170 horsepower and more. And steamers...well, they have gone the way of the dinosaurs.

## Early Tractor Competition: Caterpillar and Case

Hart-Parr may have actually produced the first gasoline-powered tractor, but dozens of other folks had the same idea. In fact, it is said that Henry Ford really wanted to produce tractors rather than automobiles. He had grown up on a Michigan farm and hated the drudgery. But his Ford automobile became so successful that the board of directors at Ford Motor Company forced him to give up his idea of producing tractors. So in the meantime, Henry was stuck building tractor prototypes for ten years before he built his Fordson tractor.

However, in California, a significant development in a unique type of traction machine was already taking place. In 1908, the Holt Manufacturing Company purchased the Best Manufacturing Company for $750,000, combining two competing manufacturers of a unique traction machine we now call a crawler.

The technology of crawlers has remained virtually unchanged since the beginning. Crawler tractors have traditionally been much more popular in the West than in the Midwest or East. That's because Western soils are often sandy and the traction of a crawler's tracks provides much better footing.

It is old, dirty, and rusty, but it still fires right up—and can still pull fifteen plow bottoms! This massive Holt Model 75 from 1925 worked in the fields near where it was built, in Stockton, California, for many decades and now resides in honored retirement on the Koster ranch near Modesto.

The crawler was a curious machine, an agricultural steam traction engine that rolled along on steel plates instead of wheels. It was developed for the sandy, marshy soils of the California Delta, where heavy steam traction engines with conventional steel wheels would sink in the swampy conditions.

Both Benjamin Holt and Daniel Best were longtime West Coast agricultural equipment developers who designed a variety of agricultural machinery: threshers, steam traction machines, and harvesters. As competitors, Holt and Best had both also started building gasoline engines at an early date. Historian Reynold Wik reports that Best was providing the gasoline engines for street-cars in San Jose, California, as early as 1892. Best tested a prototype gas tractor in 1896 and built an automobile for himself in 1898 that went twenty miles per hour (32.2kph). Fitting a gas engine into his crawler tractor was the logical next step.

Holt was also innovative, starting out in business by building harvesters in 1886. He built a crawler with unique roller plates, a machine that was dubbed a "caterpillar." His first gasoline-powered traction engine was a crawler, introduced in 1906. According to Randy Leffingwell, regular production began that year and four machines were built and sold by 1908. Three of the gas-powered crawler machines went to work on the mammoth Los Angeles Aqueduct project in southern California. The machines quickly demonstrated their value in difficult construction environments.

Holt and Best had been at each other's throats for years, spending years in litigation for patent infringement. In 1908, Daniel Best was seventy years old and tired of fighting, so he sold his business to his competition. The machine that would be the product of this merger is one of the most unique agricultural machines ever built. We know it today as the "Caterpillar." (The name "Caterpillar" was registered with the U.S. Patent Office in 1910.) The Best-Holt Caterpillar crawler tractor proved its value in three important areas of mechanical endeavor: agricultural production, heavy construction and road building, and military applications.

The Caterpillar was unsurpassed as an agricultural machine; it was a very powerful piece of equipment suited for the most difficult field conditions and very stable when operating on a hillside. It also demonstrated its capability on construction projects. Three were sold to William Mulholland for use in building the Los Angeles Aqueduct—a big job that required big horses.

Holt had already sent one of his big steam-powered machines to the City of Los Angeles for the enormous water diversion project. The efficiency of the Holt one hundred horsepower steam-powered Caterpillar was tested against traditional horse- and mule-powered equipment. The giant machine, weighing more than fourteen tons (12,712kg), quickly proved that it could outperform horses in the hot, sandy Mojave desert. The machine reportedly could haul thirty-one tons (28,148kg) of supplies a day. And it was fueled by

crude oil and required only one operator. But hauling water to a steam engine working in the desert made little sense. So Mulholland decided to try out one of the new Holt-Best gasoline-powered Caterpillars.

Extremely impressed with the performance of the Caterpillar, the City of Los Angeles bought three more right away, then purchased another twenty-five. Although the Los Angeles Aqueduct Project was eventually completed in record time (by 1913), the assessment of the tractors' performance and value seems to have been quite mixed. The desert sand was harder on the gears than anticipated, and the costs of maintenance and repairs on the machines escalated with age and use.

But the Los Angeles Aqueduct Project was an important field trial for the Caterpillar, forcing the design engineers to modify and improve their equipment. The final result was one of the most ingenious and productive little machines ever developed. The design of this machine would

Starting the big Holt engine is simple: squirt gas in the priming cups, operate the compression release lever that opens all the valves a little, and then use an iron rod to turn the big flywheel, visible on the right of the photograph. It should fire up and roar to life.

eventually be imitated and reincarnated as the Cletrac, a tiny crawler-type tractor with its own unique contribution to agricultural history.

The Caterpillar was also demonstrated to the military during World War I. Although the development of tanks and tracked armored vehicles is not a part of this tractor story, military interest in crawler technology encouraged innovations that found their way back into conventional tractors. The military development of both diesel engines and hydraulic systems for crawlers would have an important impact on agricultural tractors later on.

There is one more interesting and confusing twist to the Caterpillar tractor story. Although Daniel Best and Benjamin Holt had merged their companies in 1908, Daniel's son, Clarence Leo Best, had become disenchanted with the sale and wanted to go back into the business on his own. However, the contract agreement stated that the Bests would not go into the tractor business for ten years after the merger. C.L. Best decided to take matters into his own hands. An outstanding engineer, he quickly developed two of the finest crawler-type tractors ever built, known as the Best 30 and the Best 60.

Another serious competitor in the tractor market just after 1910 was the agricultural machinery giant, the J.I. Case Threshing Machine Company, founded in 1848. Case became involved in tractor production relatively late, considering the size of the company and the twenty years they had already put into gasoline engine research. But the Case company had an international reputation for building very high quality machinery, and they were careful to maintain their reputation with this new technology.

A large and very profitable agricultural machinery builder, the J.I Case Threshing Machine Company sold a wide range of equipment besides their famous threshers and steam traction machinery. They were known as a "full-line dealer," a company that distributed and sold a complete line of products for the farmer, even though they manufactured very few of them. By the mid-1890s, Case sold a stationary gas-powered engine known as the Raymond engine and eventually became a licensed distributor for the Raymond company.

However, Case was an international leader in the design and production of steam traction engines and threshers. The company was cautious, as building a gasoline tractor might compete with their own market for steam equipment. But by the time they finally did develop a gasoline tractor, the agricultural machinery marketplace had changed so much that the Case gas tractor complemented, rather than competed with, their product line.

The typical Case customer was a grain farmer, someone who needed a really big traction machine that could pull a four- or five-blade gang plow weighing several tons. These large heavy plows could have three, four, or five blades aligned into a single operating unit and were the most efficient way to turn over large expanses of wheatland.

In addition, the machine had to be large enough to power the largest threshing machines built by Case. The Case steam-powered machinery could handle both farm chores, and thus their gasoline tractor needed to be able to manage both as well.

Always sensitive to their customers' needs, Case was experimenting with a tractor of their own design and manufacture. But the first tractor in their catalog was not wholly their own. Case contracted with Minneapolis Steel and Machinery Company to furnish the chassis for the first Case tractors. The Case 30-60 tractor came out in 1910 and offered thirty horsepower to pull plows and sixty horsepower for the threshing chores.

It's probably no surprise that the first Case gas tractor looked a lot like the earlier Case steam engines. The Case tractor also resembled the early Twin City gas tractor, also built by Minneapolis Steel and Machinery Company. But while the design itself may not have looked innovative, Case engineering and quality experts put their machines far ahead of many others. Case had been one of the first machinery builders in North America with a materials testing laboratory. They were able to produce a product with materials of a consis-

tently high quality in a marketplace where reliability was an extremely important selling point.

A small version of the same tractor, called the Model 40, was offered in 1912, followed the next year by an even smaller gas-powered tractor, the Model 12-25. This last little tractor was very popular, and Case eventually sold more than thirty-three hundred units during the five years it was produced.

## Early Leaders in the Industry: Avery, Heider, and Wallis

There were scores of small tractor builders by the end of the first decade, but only a few of them offered important innovations that gave them a special niche in the history of tractor technology. By the end of the first decade, many old-line agricultural equipment dealers were thinking of adding a gas tractor to their product line. Farmers were asking their local dealers about tractors, and there were numerous articles about power farming in the farm papers.

By 1907, the Hart-Parr Company had sold hundreds of tractors. Other agricultural equipment giants, the huge conglomerates such as the newly formed International Harvester Company and the long-lived J.I. Case Company, were keeping a watchful eye on tractor development. Henry Ford was building prototypes, experimenting with all sorts of designs between 1906 and 1916. Agricultural machinery production was headed in a new direction, and that direction was pretty clear. The marketplace wanted tractors.

Dozens of auto builders were looking at the Hart-Parr machines and getting into tractor production themselves. The tractor manufacturing industry would expand enormously in the second decade, including automobile manufacturers as well as barnyard tinkers and shade-tree engineers. The Model T Ford was introduced in 1903, and it didn't take long for hundreds of amateur mechanics to begin modifying the automobile for farmwork.

The old-line companies had a dilemma: should they dig into their own pocketbooks and finance

tractor design and development, or should they find some enterprising fellow who had already developed a working prototype and buy his design? There would be as many solutions to this problem as there were agricultural equipment companies. The Avery Company of Peoria, Illinois, built its own tractor from the ground up. The Heider brothers of Carroll, Iowa, would design a tractor made from premanufactured parts, and H.M. Wallis of the J.I. Case Plow Works would buy a complete prototype outright. History demonstrates there was no optimal alternative.

## Avery: Imitation Over Innovation

The Avery Company has now faded from sight, but for many years the Bulldog machines from the Avery Company had a large and loyal group of Midwestern customers. Now they have a select group of devoted collectors.

Starting as an agricultural implement dealership in 1874, the Avery Planter Company began building steam traction machines and threshers in 1891. The Avery brothers, Robert and Cyrus, at one time built six sizes of steam traction machines, with the Avery Bulldog as their mascot and trademark. The company was very competitive, matching product lines with many other industry giants such as the J.I. Case Threshing Machine Company.

In 1911, after Hart-Parr gas tractors had begun making a significant dent in steam engine sales, the Avery Company put together its own gasoline-powered model. They started out building big square tractors that resembled the Avery steam engine line. With a tall, squared-off canopy over the operator, steel driving wheels that were over six feet high, and a tall smokestack, the Avery Model 40-80 gas tractor looked like a steam traction engine.

Then in 1913, after the popularity of a little machine from the Bull Tractor Company and the excitement over the Hart-Parr "Little Devil" model, Avery quickly changed and started building small tractors and three-wheeled tricycle

tractor models. They experimented with a motor-cultivator during the late teens and even produced a crawler-type tractor in the early twenties. But much of their product line was developed later than competing models; their strategy of imitation rather than innovation meant that they were always left behind.

The Avery tractors did have their moment in the limelight, though. An Avery tractor was a big attraction at the famous annual tractor show held in Fremont, Nebraska. The smallest tractor at the 1916 show was the Avery 5-10. It was an important machine because it demonstrated that a small

functional machine could actually be built. And it demonstrated that the marketplace was actually interested in buying such a small machine.

The Avery 25-50 was one of their most popular models in what was then classified as the midsized range. It was introduced around 1914 and continued to sell until around 1927. It was a state-of-the-art machine when it was first introduced but became hopelessly outdated by the end of its production run a decade later.

Some folks say that Avery innovation fell behind in the World War I years because the corporation refused to bid for government contracts;

**OPPOSITE:** Picking up bundles of shocked wheat for threshing, Waverly, Nebraska.
**ABOVE:** Avery tractors used oil instead of water as a coolant, a system that was popular with several other makers during the early years of the industry. This Avery 25-50 is part of Doug Peltzer's extensive collection and is in essentially original, unrestored condition.

others criticize the old-fashioned machinery they were building. In either case, Avery was pushed into bankruptcy in the 1920s, reorganized, and continued to hang on only until World War II.

## 1911: Heider Model A

Iowa has an extraordinary tradition of tractor development; the Hart-Parr and the Waterloo Boy are two of the most famous early tractors from Iowa. The list also includes the Heider tractor, built by John and Henry Heider, who moved their business to Carroll, Iowa, around 1904.

John and Henry Heider formed a business in 1903, selling replacement parts such as wagon

trees for farm equipment. Henry had already been awarded a patent for his improvements on an evener, sometimes called a draft equalizer, a device that "evened" the load a team of draft horses was pulling. The Heider four-horse evener was sold through jobbers across the Midwest and the business began to grow. John Heider was an accountant, and together the brothers made a good business team.

A history of the Heider Manufacturing Company by Dave Erb and Eldon Brumbaugh in

their book, *Full Steam Ahead*, says that Henry Heider was a genius at taking standard parts from various sources and adapting them to his own designs. His tractor designs would make full use of this particular skill.

As the business expanded, Henry saw the need for a small tractor, one the average farmer could afford. He began assembling a prototype around 1908, using various standard parts such as wheels, chain drives, and even a manufactured engine, all bought from other suppliers. The first Heider tractor was shown to the public around 1910. According to R.E. Hoffman, writing in the July/August 1994 issue of *Antique Power Magazine*, it is estimated that approximately thirty-seven of the Heider Model A tractors were built and sold.

The first Heider tractor featured a unique galvanized metal radiator. There may be only one left in existence, a rare Model A pulled from an Atlantic, Iowa, fence row by Dave Soper. Soper rescued it in 1962 from the field where it had been sitting since 1924. He took years to carefully research the machine, finally completing his restoration of the Heider Model A in 1993.

The Heider Model A was quickly followed by a Model B, introduced in 1911. The Heider brothers had to enlarge their production facilities and hire additional workers to keep up with demand. It is estimated that about seventy-five Model B tractors were built before the Heider brothers ran into difficulty. They were primarily in the implement sales business, not in the tractor manufacturing business, and thus were not equipped to meet the demand for their machines. The Rock Island Plow Company came to their rescue, contracting to distribute and sell the Heider tractor through the branch houses of the Rock Island company.

Model C made its debut in 1914, an immediate success selling for only $995. With a Waukesha engine and seven forward speeds, it was in great demand, and its popularity put the Heider brothers in another crisis. They simply did not have the capacity in their little factory to fill their orders. Their solution was to sell the Heider tractor

All sorts of technologies were cobbled together in the early days of tractor design in an effort to make the basic machine efficient and economical in its specialized job. This Rock Island Heider, built during the last year of World War I, uses an innovative—and not very durable—friction drive mechanism.

division to the Rock Island Plow Company, which immediately moved manufacturing to a much larger facility across the river in Rock Island, Illinois.

The Heider Manufacturing Company continued to build eveners, ladders, and agricultural specialties, expanding their original business. Henry Heider contracted to continue working with Rock Island Plow as a consultant, designing tractors for a retainer of $10,000 a year. He finally resigned as consultant in 1922.

Tractors with the Heider name continued in the Rock Island Plow Company catalog until

1927, when a similar machine started being sold under the Rock Island moniker. The Heider tractors had made an important contribution to tractor manufacturing because the Heider brothers had demonstrated that it was possible to design and sell a machine made up of previously manufactured and conscientiously assembled components. By selecting their suppliers carefully, the Heiders were able to produce a reliable, affordable machine, which helped them compete successfully with industry giants such as Case and John Deere.

## 1913: Wallis Cub

The Wallis Cub, produced by J.I. Case Plow Works, is the tractor that should have taken over the world. Just over a decade after Hart-Parr introduced their first production machines, the industry had already matured enough to produce the remarkable Wallis tractor. It had superior engineering, remarkable reliability, and all the distribution and sales support it needed. It featured a four-cylinder engine, one of the first tractors to do so. And it was one of the first tractors that looked like a tractor rather than a locomotive. It was a great little machine.

Its compact appearance resulted from the introduction of the unit frame, a one-piece tractor frame formed of boiler plate. This plate served the dual function of frame and crankcase. Other tractor builders of the time were building a frame out of channel iron, then mounting the engine, transmission, and wheels to the frame. The Wallis tractor created a new concept in tractor chassis construction, creating a much lighter and, therefore, much more efficient machine. Although the Fordson tractor of 1918 is sometimes touted as the first unit-frame design, the concept was introduced in the Wallis Cub, five years earlier.

The Wallis Cub was a dependable little machine, and its marketing staff had a flair for promotion. The Cub demonstrated its performance and reliability in a run from its home plant in Cleveland, Ohio, to the annual tractor show in Fremont, Nebraska. The trip earned it the nickname "The Thousand Mile Cub," a slogan that was prominently displayed in advertising and promotional material.

This machine also had an impressive pedigree. The genius behind its introduction was Henry M. Wallis, a son-in-law of J.I. Case and then the president of J.I. Case Plow Works. Wallis, looking for a solid little machine to add to his agricultural equipment line, acquired the tractor design from inventor Robert Hendrickson. Wallis knew the agricultural machinery business and Hendrickson had designed a terrific little tractor that needed a sales and distribution network. It was a good match.

So why wasn't this little machine the biggest-selling tractor of its time? Why isn't the company around today? Why have few people outside of the tractor enthusiast world heard of the Wallis? If you guessed that it might have something to do with internal corporate competition between the Case tractor line and the Wallis tractor line, you'd be on the right track. The Wallis Cub would ultimately be a casualty of a curious competition between the J.I. Case Threshing Machine Company and J.I. Case Plow Works.

It's a strange story in tractor history and an odd predicament for any business. For more than fifty years, there were two completely separate companies, both founded by Jerome I. Case and both headquartered in Racine, Wisconsin. They both used the famous Jay Eye Cee trotting horse in their advertising campaigns.

Members of the Case family sat on the boards of both companies, and for many years Jackson I. Case, the only son of the founder, was the president of the Plow Works. It should be no surprise to discover that it finally took a Wisconsin Supreme Court decision in 1915 to straighten out the mess.

Both Case companies had a gas tractor in their respective catalogues, and the marketplace was pretty lively in 1915. Although the innovative Wallis tractor line stood up well against other machines, the Plow Works company itself, a full-line manufacturer and distributor of agricultural equipment, would not be as successful.

Wallis tractors continued to be innovative machines. The Model K, offered in 1919, featured completely closed gearing in a continuous oil bath. New Wallis models were designed and introduced until the late 1920s. But weakened by litigation and tension with the other Case company and overwhelmed by larger manufacturers, the company could not survive. J.I. Case Plow Works was acquired in 1928 by the Canadian agricultural manufacturer Massey-Harris. The Wallis tractor faded into the Massey-Harris tractor line and eventually disappeared altogether around 1938.

# Chapter 2

# Cultivating the Market— Power Farming

## *(1912 to 1920)*

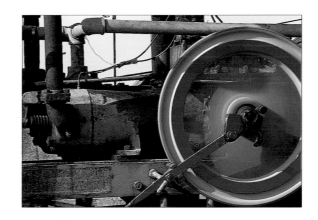

Early in the second decade of the twentieth century, a little tractor appeared that finally took into consideration the needs of the average farmer, the homesteader with a 160-acre (64ha) farm. The early gas-powered tractors had resembled the old steam traction machines. Looking like derailed locomotives, the early machines weighed several tons and were nearly as difficult to drive as the steam machinery they were replacing. And they cost a fortune, far beyond the pocketbook of most farmers.

But one fine summer day in 1913, a little tricycle tractor appeared at a tractor demonstration. It was so popular with the public, and so afford-

able, that all manufacturers hurried back to their engineering departments to design something similar. It was called the Bull, built by the Bull Tractor Company of Minneapolis, Minnesota. Tractor building would now be heading in a new direction, designed for the average farmer, the fellow with the small farm.

## Selling Tractors—Tractor Shows, Demonstrations, and Tests

There are several landmark shows and expositions in tractor history, gatherings where tractors were rolled out to perform for enormous crowds. They became watershed events because the performances at some of the shows provided inventors with new ideas. Inspired by the exhibits, some engineers immediately rushed back to their drawing boards. There is one story that the Hart-Parr "Little Devil" was the product of just such inspiration. By the time the engineers had returned to the Hart-Parr headquarters in Charles City, they had reportedly worked out all the details in their heads.

In those days before radio and television, state and county fairs and agricultural shows were typically the only public venues for tractor sales. Of course, there was print advertising in magazines and newspapers, but this had a limited audience. The fairs were weeklong social events with all sorts of livestock competitions as well as tractor and equipment demonstrations. Plowing competitions were frequently a featured event. Horse teams had their pulling and performance contests, and so did the big traction machines. State and county fairs continue to be important annual events today in agricultural America. They are still good places to check out the new tractor models and watch the pulling contests on the old ones.

A few machinery shows were especially significant. The Centennial Exposition, one of the earliest important exhibitions of American agricultural machinery, was held in Philadelphia in 1876. The coast-to-coast railroad had been completed just a few years earlier, in 1869, so it was now possible to ship heavy equipment from the other side of the country for the show. The J.I. Case steam machine from Racine, Wisconsin, was demonstrated, as were the binders and reapers built by The McCormick Harvesting Machinery Company of Chicago. The Centennial Exposition

THE NEW
CASE
TRACTORS

J. I. CASE CO.
Established 1842 • • Incorporated
RACINE · WISCONSIN · U.S.A.

OPPOSITE: While some tractor designers of the very early years stressed light weight and simplicity, others essentially duplicated the massive bulk and weight of steamers. This cockpit shot of a beautifully restored Minnesota gasoline tractor is a case in point, with the boiler replaced by an equally hot exhaust manifold.

LEFT: By 1920 the Case company had pretty well worked out the details of what was required for an efficient tractor and announced a new line of designs with this catalogue.

was the first national event that demonstrated the steam power that was now readily available to the farmer.

Another important event frequently mentioned in farm literature was the Winnipeg Motor Contest, a Canadian event that originated in 1908. This contest was among the earliest large events in which gasoline engines were pitted against steam traction engines, with careful records kept regarding fuel consumption, horsepower, and performance.

Started simply as a public tractor show, the exhibition became a contest the following year when manufacturers were allowed to enter and demonstrate steam tractors. This, of course, meant that in addition to the general public, the audience included salespeople and tractor designers. The results of the contest were widely published in the agricultural newspapers as well as in the sales literature of the various companies whose machines made a good showing. For the first time, farmers were able to evaluate and compare machinery in the field. Case tractors came home from the last Winnipeg show, held in 1913, with four gold medals and a silver trophy.

Another watershed tractor event was the National Farm Tractor Show held in Fremont, Nebraska. An annual event, the show held in 1916 was especially memorable. Unconcerned with the fact that Europe was at war, tens of thousands of Midwesterners traveled the country roads to Fremont for a tractor demonstration. The attendance estimates are staggering. The Fremont show was impressive not only because of the tractor performances but also because of the estimated fifty thousand participants.

A reporter for *The Country Gentleman*, then the largest and most respected agricultural magazine in America, wrote:

> *A statistician of the Fremont Board of Trade figured that, sure enough, Nebraska farmers burned up half-a-million gallons of gasoline getting to the show and home again. The Fremont Candy Kitchen made a million sandwiches or thereabouts. A Fremont traffic constable developed paralysis of the right arm directing traffic. Between four and five thousand visiting automobiles streamed through Main Street and thence on out Lincoln Highway to the tractor grounds on the biggest day of the show. The cars were parked along a mile and a half of roadway, from five to fifty deep. It was on Wednesday, August ninth that the local papers counted 50,000 folks as holiday visitors to their trim little city and its prairie suburbs. Fifty*

Irv Baker's even bigger, even louder 14-28 Rumely OilPull. Tractors at the time were rated (often optimistically) by horsepower produced at the drawbar (14hp in this case) and at the pulley (28hp for this example), the difference being the result of losses in the drive train.

*thousand "tractor fans" may be an overcount of a few battalions, but as there were no turnstiles to check up by, why not give the local optimists all the benefit of the doubt.*

Henry Ford was present at the 1916 Fremont show, and reporters followed him around, asking about the new tractor that he was rumored to be developing. Henry was tight-lipped about design details but did comment about the concept of power farming and the future of the tractor. After the Fremont show it was clear to everyone that there was no turning back—like it or not, the tractor would have an important role in farming.

One more important event to measure tractor performance was soon to become an institution: the Nebraska Tractor Tests. A Nebraska farmer and politician, Wilmot F. Crozier, had purchased a tractor, only to discover that it was a lemon. He replaced it with a Rumely OilPull tractor, built by the Advance-Rumely Thresher Company of LaPorte, Indiana. To his surprise, he found it surpassed the manufacturer's stated specifications. He wondered if there was a way to uniformly test the performance of all tractors so that consumers would be able to evaluate the results.

A bill was introduced and quickly passed in the Nebraska legislature in 1919 to require all tractor manufacturers who wanted to sell machines in the state of Nebraska to submit them to standardized tests. A test laboratory was set up at the agricultural college of the University of Nebraska in Lincoln. Tractor builders began submitting machines for testing the following year, 1920.

The first machine to undergo rigorous evaluation was a Waterloo Boy Model N, marketed and distributed by the John Deere Company. The next six machines to be tested were all submitted by the J.I. Case Threshing Machine Company.

Through the years, the laboratory in Lincoln has received and evaluated many of the historic tractors. The first Cletrac model from the Cleveland Tractor Company, the little 12-20 Model W, was put through its paces in July 1920. The first diesels from Caterpillar were tested there, starting in 1932. The beautifully styled Oliver Hart-Parr Rowcrop 70 got its examination and a passing grade in 1936. The Case 500 diesel with power steering gave it the old college try in October 1953. Testing continues there to this day, and over the years the Nebraska Tractor Tests have faithfully recorded engineering milestones in agricultural development.

Competition improved tractor design, but so did intellectual collaboration. When two professional engineering societies, the Society of Tractor Engineers and the Society of Automotive Engineers, merged in 1917, the quality of tractor engineering immediately improved. As engineers traded ideas, there was a noticeable improvement in both the design and construction of engines, transmissions, and other mechanical parts. The techniques of fabrication also improved, inspired in great measure by contributions from engineers from the automobile industry. The combination of shows, competitions, testing requirements, and engineering collaboration during the second decade of the century resulted in a tremendous burst of mechanical creativity. Tractor design began to take quantum leaps forward.

## 1914: Allis 10-18

The Allis-Chalmers Company first entered the tractor market with a tricycle tractor in 1914. They too had seen the Bull tractor demonstration and decided to build a small machine. But the

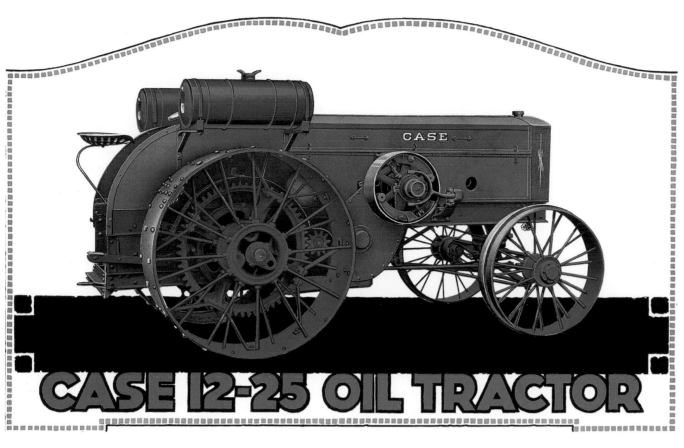

A catalogue illustration of Case's radical new 12-25, the racing car of the back forty. An expensive, well-engineered design, the 12-25 was a decade ahead of its time and competition.

## CASE 12-25 OIL TRACTOR

Allis-Chalmers 10-18 proved to be just the beginning of a long and profitable line of gas-powered tractors from this company. Continuing with a plow/cultivator in 1919 and their four-wheeled Model 18-30 the same year, Allis-Chalmers tractors soon had a reputation for being as solid as the other machinery the company built.

The Allis-Chalmers Company, a conglomerate of four diverse businesses based in Milwaukee, Wisconsin, was one of the first partnerships to bring a really workable tricycle tractor into production. The 10-18 was an innovative design, a little machine capable of pulling three plows. Although we now know it was a sturdy and reliable performer, it did not attract many customers or much interest at the time.

The Allis-Chalmers Company had been formed around 1901, when the E.P. Allis Company merged with Fraser & Chalmers of Chicago, the Gates Iron Works, also from Chicago, and the Dickson Manufacturing Company of Pennsylvania. While many other tractor builders had an agricultural equipment company or two in their lineage, Allis-Chalmers was descended from steam turbine pump builders and electrical generator manufacturers. This diversity in product lines may explain how they managed to stay profitable for many years when conventional tractor builders were suffering from a glut of tractors on the market.

While J.I. Case and John Deere were each introducing a wide assortment of tractors to the farmer, Allis was offering only one or two. But the Allis company had a strong national reputation, and Allis engines were noted for their excellent design. The tractors themselves were solidly built. The advertising for the Allis-Chalmers Model 10-18 declared "It is the only tractor with a one-piece steel, heat-treated frame—the only tractor frame with no rivets to work loose—that cannot sag under the heaviest strain."

Allis also assured their products' superiority and longevity by building a better engine for their tractors. They already had a well-earned reputation for sixty years of superior quality machinery when they introduced the 10-18. Highly prized by collectors, there may be only a dozen Allis-Chalmers 10-18 machines still remaining of the twenty-seven hundred that were built between 1914 and 1921.

## 1914: Case 12-25

The Case 12-25 is the tractor that was on the drawing board of the J.I. Case Threshing Machine Company when the little Bull first appeared. The Case 12-25 was ahead of its time for 1913: it was the first tractor with a roller-chain and sprocket drive that was virtually impossible to break, the first with a closed transmission, and the first to be designed specifically for the average farmer.

While the successful Bull tractor had somehow wandered into the small tractor marketplace and been lucky, the Case corporation did a great deal more analysis before building any new product. Other tractors typically had a little galvanized

roof over the operator; the seat on the new Case 12-25 machine was open to the elements. And while other tractors had all of their components exposed, this machine looked more like Barney Oldfield's racing car: all moving parts except the wheels were out of sight, tucked under sheet metal covers that enclosed the entire machine. It was designed to do the regular farm chores, rated to pull four plows with its twenty-five horsepower engine, and its look was sleek and tidy.

The Case 12-25 sold for $1,350, a princely sum compared to the $335 price asked for the Bull, but Case machinery was always known to be comparatively expensive. The quality was commensurate with the price. Just as the 12-25 was being built, the Case company rushed to build the Case 10-20, a tricycle tractor to match the Bull. Fortunately, the development of the 12-25 would put their engineers ahead in the race for a tricycle model. The new 10-20 machine turned out to be

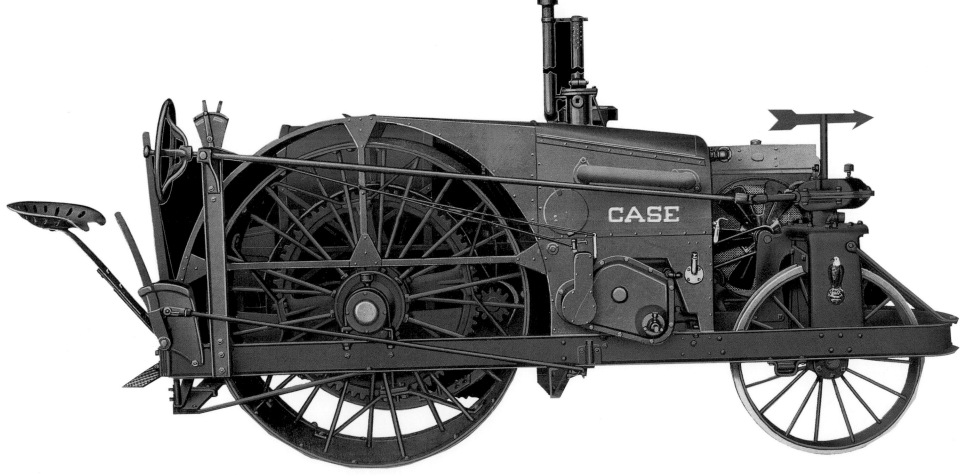

more popular than both the Bull and its older brother, the 12-25. The little Case tricycle would sell more than sixty-six hundred machines over its four-year production life, even at Case prices.

## Case 10-20: The Case Tricycle

While the first-generation tractors were big and heavy, looking like the steam-powered monstrosities that preceded them, the machines of the second generation of tractors were much lighter, bearing a greater resemblance to a modified Model T Ford. But the transition was not made in one single great leap. The period between the two was a curious time when many tractor builders experimented with the tricycle tractor. Today the tricycle configuration for tractors, with one wheel up front and two behind, is very common, especially for row-crop farming. But back then it was an innovation that still had some substantial engineering problems to overcome.

The Model 10-20 was the first small tractor that Case built; the company put it on the market in 1915 as a response to the popularity of the Bull. Like the other tricycles, it had some engineering flaws. Critics have said that it was top heavy, with the weight distributed too far forward, and that provided poor field visibility for the operator. But overall, it was probably the best-engineered tractor by any manufacturer when it appeared, and the engine design alone would prove to have a long and useful life.

Case was one of the few tractor builders that designed and made its own engines. The 10-20 would be the very first Case tractor with a four-cylinder engine. They designed an overhead-valve-type engine, with a cylinder block that sat on top of the crankcase. It was a rugged and reliable design that Case would use for the engines of all of their tractor models until 1929.

The 10-20's popularity demonstrated that there was serious interest in a small tractor. The popularity also helped prepare customers for the introduction of the lightweight Case Model 9-18, a small, four-wheeled machine introduced by Case in 1916.

## 1914: Best Crawlers

One more example of the extraordinary change in the direction of tractor technology was introduced at the beginning of World War I: the Best crawler. And yes, the Best was best. Under the terms of the contract worked out by Daniel Best and the Holt brothers in 1908, the Best family was

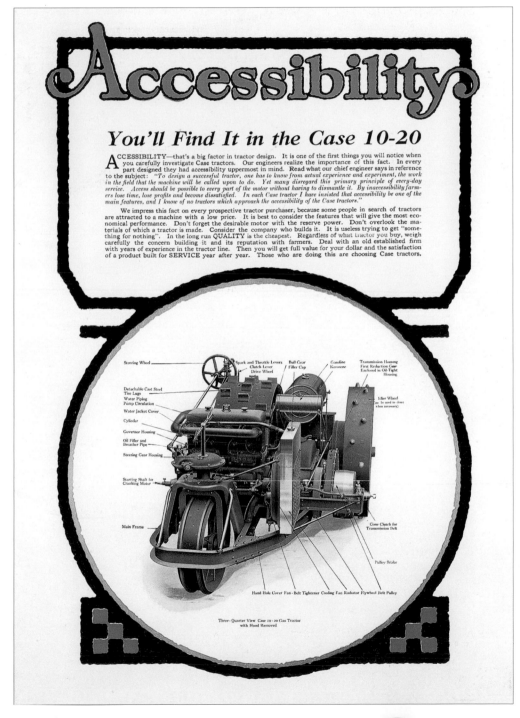

supposed to stay out of the crawler manufacturing business for ten years. But Daniel's son, Clarence Leo Best, couldn't wait and by 1910 he was trying to get out of the contract.

An extremely innovative engineer and a maverick at heart, Clarence Leo Best took matters into his own hands and started a new crawler business around 1912, in complete contempt of the contract. Needless to say, he was sued by the Holt brothers almost immediately. However, he continued to design, build, and sell crawler-type tractors, defiant of pending litigation.

Best could not use the name Caterpillar (since it had already been trademarked by the Holt brothers), so he called his machines "Tracklayers." The first machine out of his new shop was a tricycle. It's difficult to envision the tricycle and crawler as a workable combination, but the Best Model 75 had two crawler treads for rear wheels and a single conventional wheel up front. While other tricycle tractors were designed for the small farm, this machine reportedly weighed seventeen tons (15,436kg) and sold for nearly $6,000.

His next crawler was one of the most innovative agricultural machines ever built: the Best Model 30, nicknamed the "Muley." Extremely advanced for its time, the 1914 Muley had more in common with the technology to come in 1920. In fact, many of the major elements of this design have remained unchanged to the present day. The Best 30 was a crawler tractor without any wheels; it was steered by tillers and ran only on treads, just like the modern crawler machines. A smaller version known as the 8-16 was introduced in 1915, and the famous Best Model 60, the larger version, was introduced in 1919.

Litigation was a lifestyle for the Best and Holt companies, and it is estimated that between 1907 and 1918 the two groups spent $1.5 million suing each other. The two had first merged in 1908 when old Daniel Best sold his company to the Holt brothers. The two companies would merge again in 1925, after a second buyout, which caused the extraordinary Tracklayer technology to become part of Caterpillar.

# 1916: Cletrac Model R

The Cletrac was also a crawler, a direct copy of the machine designed by Daniel Best. It was manufactured just outside of Cleveland, Ohio, and was the product of one of the strangest episodes of manufacturing and marketing chicanery in the tractor industry. The story of how a California tractor came to be manufactured in a Cleveland suburb is an interesting saga.

Crawlers, those curious tractors that run on treads rather than wheels, were developed in California for local farms. Designed for the soft soil conditions in the Sacramento Delta farmland, these funny-looking machines were sold primarily on the West Coast. But they had demonstrated their power and reliability on construction projects as well on the delta farmlands.

Crawlers have a very low center of gravity, making them very stable for work on hillsides where wheeled tractors would tip and roll. They are also able to pull tremendous weight for their size; for instance, one two-ton (1,814kg) crawler model can typically pull over thirty-two hundred pounds (1,453kg). This makes it an extraordinary machine for heavy chores like house-moving and logging.

Demonstrating their machines at every opportunity, the C.L. Best Company of San Leandro, California, exhibited their Best crawlers at the Pan-Pacific International Exposition in San Francisco in 1915. Visiting businessman Rollin White saw the Best crawlers while touring the exhibition. A wealthy engineer, White was interested in expanding the investments of his family corporation. The White Company of Cleveland, Ohio, founded by Thomas White in 1859, first built sewing machines, then roller skates and bicycles. By 1901, Thomas White's four sons were building steam-driven trucks and vans. The White enterprises were noted for their innovative machinery.

Seeing the crawler, Rollin White understood immediately that its business potential was tremendous and had the capital to invest in a new factory. According to historian Randy Leffingwell

in his book *Caterpillar*, Rollin White sent his sales manager, C.L. Hawkins, to talk with the Best sales manager about licensing the design and building a crawler for use in the Midwest. The Best Company was not interested in such an arrangement.

In those days, however, the Best Company was always short of operating cash. To make ends meet, they sold their experimental models whenever possible. They reportedly sold one of the demo models, the little Best 8-16, from the floor at the exposition. It was quickly crated and sent back to Cleveland, where it was turned over to White's design engineers. C.L. Hawkins also bought enough stock to control the Best Company.

Copying, stealing, reverse engineering, modifying—call it what you like. Perhaps the Whites felt that owning controlling interest in the company justified filching the design. After evaluating the model, the White brothers did make an important improvement, a modification to the Best design. Clarence White was an agricultural engineer, and he had been impressed with the performance of another California tractor known as the Yuba Ball Tread Tracklayer. Remembering the Yuba design,

the White brothers decided to redesign the tracks on the Best 8-16, omitting the roller balls and the ball race that contained them. With this modification, the Cletrac was born.

According to Leffingwell, the Whites quickly built a new manufacturing plant in the Cleveland suburb of Euclid, Ohio, and put together an advertising brochure. They photographed the little modified Best 8-16 prototype they had bought in California for their brochure, named their new creation the Cletrac, and sent the pictures out to dealers around the world. Twelve thousand tractors were presold.

Clarence White also improved and patented the steering system on the Cletrac machines, starting with the Model R. He laid out a controlled differential steering mechanism for clutching and declutching, and he called his improvement "Tru-Traction." But the basic Best crawler design remained the same. Cletrac company literature stated that although the Cletrac Tractor Company had made many improvements, the first machine had been so satisfactory that no basic change to the fundamental design had ever been made.

A good-as-new little Cletrac from about the time of World War I, circa 1916. This one has been restored by Bill Bechtold and his family, among the leading devotees of Oliver and Cletrac machines.

## 1916: LaCrosse Happy Farmer

The rage for tricycle tractors was in full force when the Happy Farmer, built by the LaCrosse Tractor Company of LaCrosse, Wisconsin, was introduced in 1916. Historian C.H. Wendel reports that the company produced tractors for only about six years, going out of business when it was absorbed by the Oshkosh Tractor Company. Most of their model offerings were three-wheeled tractors. The industry quickly abandoned tricycle tractors when they proved to be less maneuverable as well as less stable than the four-wheeled variety.

There are mixed reports about the popularity of the machine. Some sources say the customers were delighted and that the Happy Farmer advertising made a positive impression on farmers. It was a catchy company name. There are others who comment that there was once a saying about the owner of a Happy Farmer machine: "He wasn't happy for long."

Happy Farmer tricycles were built in at least four models between 1916 and 1920. The company also built a singular four-wheeled model known to tractor enthusiasts as a line-drive tractor. This machine had no steering wheel; instead, it was fitted up with a set of leather reins so the owner could operate the tractor as if driving a team of

horses. The Happy Farmer remains a memorable
tractor today, special to collectors for its amusing
name and rarity rather than for its engineering.

## Tractors in the World Marketplace

Before World War I, American steam traction
engines and agricultural machinery of all types
were exported to customers in Canada, Europe,
England, and South America. World War I began
in Europe in June 1914, but America did not enter
to assist its allies until 1917. Although many
countries had their own agricultural machinery
industries, the combination of lower prices and
technical superiority made American machines
more attractive. After the beginning of the war,
both manufacturing and agriculture in Europe
were greatly disrupted, making American exports
essential to maintain European agriculture.

Farm machinery export has a long history.
Historian Robert Pripps notes that both the
Advance-Rumely Company of LaPorte, Indiana,
and the Hart-Parr Company of Charles City,
Iowa, sent trainloads of their tractors to western
Canada years before the war. Caterpillar sales
records show that more than one hundred
machines were shipped to Buenos Aires by the
end of 1911. The J.I. Case Company received an
order from the Russian government for one hundred of its twenty-five horsepower steam traction
engines in 1900 (they had been exporting steam-
powered threshing rigs to Europe and South
America since the 1880s.) The history of the Case
company by David Erb and Eldon Brumbaugh,
*Full Steam Ahead*, reports expanding the sales territories to South America in 1885 and opening a
Case branch office in Buenos Aires in 1890.

Case was not the only manufacturer to do
well abroad. The outstanding Hart-Parr machines
were also widely admired in Europe and sold well
there. Unfortunately, the popularity of Hart-Parr
machines overseas would put the company in
jeopardy. At the outbreak of World War I in 1914,
Hart-Parr had shipped a substantial number of

machines to France and Germany. Hart-Parr had
no way of getting paid when the war broke out.

The financial distress that these European
shipments caused the corporation in Charles City
eventually cost Charles Hart and Charles Parr
their positions. Defaulted payments combined
with the cancellation of a lucrative contract to
manufacture World War I armament put the
Hart-Parr Company in dire economic straits. A
disgruntled executive board voted Hart and Parr
out of the active management of their own
company in 1918.

If European sales were the downfall of Hart-
Parr, they proved to be a windfall for Henry Ford

and Son, Inc. and their new Fordson Model F. There are probably more Fordson tractors in the United Kingdom and Europe than there are in America. Always opinionated about national issues, Henry Ford protested American involvement in World War I. An outspoken pacifist, Ford sent two prototype Fordsons to the British Board of Agriculture for testing in 1917. War-ravaged England was desperate to find a small, cheap tractor to help quickly remedy or at least alleviate the impending food shortage.

The Fordson proved to be just the machine they needed. Small, inexpensive, and easy to mass produce, this tractor was uniquely suited to help

resolve the British crisis. Henry Ford generously donated the European patent rights and agreed to build a Fordson factory in Cork, Ireland. The Ford family had emigrated from County Cork, and the donation was made to honor the Ford heritage.

Although British tractor builders were understandably upset, the Fordson tractors seemed to be the best solution to the dilemma. Until the new factory in Cork could begin production, however, Henry quickly shipped six thousand of his American-built Fordson machines to England. In the end, manufacturing of Fordsons in England would continue until 1946, long after Fordson production in America had come to a halt.

**It is hard to improve on perfection, so the basic design of the Cletrac has stayed the same for decades. Another model from the Bechtold collection, the tracked crawler is an extremely useful machine for orchard and vineyard work.**

An early John Deere Model D "spoker" on steel wheels, the most durable tractor design of all time. When this machine was built in 1925, Deere was just getting into a tractor market already covered by Fordson, Case, and many other manufacturers.

The John Deere Company also went to Europe and South America early in the century. The company exported tons of agricultural machinery to overseas markets. During the early 1920s, when agricultural machinery markets were generally depressed in America, the biggest export market for the Deere Model D was in Argentina. Introduced in 1924, the John Deere Model D did well in Russia, too; these sales kept the company afloat during tough times at home.

The tractor industry would be profoundly influenced by international trade. Within less than fifteen years, the American tractor industry had grown to be an international business. Foreign sales would help keep some agricultural equipment companies solvent. And it would be the inventive-

ness of a French tractor designer modifying an exported McCormick binder that would inspire an important, entirely new device. The power take-off, or PTO, made its American appearance in 1918, modified and adapted by International Harvester engineer Edward A. Johnston. This device would revolutionize the industry.

## 1917: International Harvester 8-16

The power take-off device, or PTO, as it is commonly known, first appeared in 1917 on a small tractor developed by International Harvester. This small tractor had been developed by I-H after carefully watching the sales of models from two other very successful builders, the little Case 9-18 and the new Fordson. The success of these two machines convinced the International Harvester Company that they also needed to offer a smaller machine, and their engineering department quickly came up with one. After all, I-H had been in the tractor business for years, selling both the Titan and Mogul machines. The new I-H Model 8-16 would take the best that International Harvester had to offer and add something even better.

The power take-off is a splined shaft protruding at the rear of the tractor, under the operator seat and is part of the coupling system used when attaching the tractor to a binder or mower. This shaft made it possible to use power from the motor of the tractor to provide motive power to an implement. And for the first time, a farmer could operate both pieces of equipment from the seat of his tractor. Many farm chores now became one-person operations.

The most important farm chore that the model 8-16 took on was operating a binder. It was now possible to cut and bundle a crop by towing and operating the binder using the power from the tractor motor. The combination was unbeatable and quickly became a very popular team. International Harvester would sell more than eight thousand Model 8-16 tractors before it redesigned the machine in 1922.

# 1919: Case Crossmotors

One of the most popular and durable early tractors was the Case crossmotor series, a tractor whose engine was designed to sit sideways within its frame. There is something very appealing about the solid little bulldog appearance of the Case crossmotors. Their jaunty exhaust pipe is also a memorable feature.

By the end of World War I, there were an estimated two hundred tractor manufacturers in North America. Historian C.L. Wendel lists forty-five kits that were available to turn the Model T Ford into a tractor as needed. There seemed to be many choices available to someone who wanted to acquire a tractor.

Why set the engine crossways? Was it a gimmick, something to set Case tractors apart from the competition? There were several good reasons for this radical departure from the convention of the time, all driven by engineering considerations.

First, the farmer usually had to be his own mechanic, maintaining and making simple repairs on his own machine. He wanted and needed a tractor that was easy and straightforward to fix. Crossmounting made the engine much more accessible for routine maintenance. Second, crossmounting allowed Case to install a fully enclosed two-speed transmission with straight-spur gearing, making fewer parts to break. Third, threshers, grinders, mills, and other equipment could be run directly off the end of the crankshaft, now easily accessible on the left side of the tractor. Finally, putting the engine in the middle balanced the weight, making the machine safer. It solved the problem of instability and made it difficult for the tractor to roll over.

The best of the four crossmotor tractors built by Case was the Model 15-17. Introduced in 1919, it proved to be the most popular tractor in the series, with sales exceeding seventeen thousand units in its five years of production. It featured a tough, four-cylinder engine (the John Deere was using a two-cylinder engine) and two other features that provided greatly improved reliability for the kinds of field conditions that Case customers faced.

Dust and chaff in the carburetor were now minimized, thanks to an air-intake system with an innovative water filter. Hot exhaust gases from the engine were routed around the intake manifold, warming the intake air. This kept the engine happy and running well during those cold Wisconsin winters, as well as improving fuel economy.

**A Case 12-20 plowing, sometime in the mid-1920s. The tractor's ability to plow continuously, without the frequent rest breaks required by horses, was one of the selling points for the new technology.**

The traditional customers for Case equipment—threshers and traction engines—were the wheat farmers in the northern Great Plains. These farmers needed a machine with enough horsepower to pull the heavy gang plows with four or five blades and enough muscle to power the thresher when needed. They needed a machine that they could count on to perform in the cold weather of the northern prairie. Reports say that many of the customers for this particular machine were first-time tractor buyers. The new Case crossmotor more than filled the bill.

## 1917: The Fordson

If you wanted to list the five most important tractors of the last hundred years, the Fordson would have to be near the top of the list. Henry Ford not only introduced the first mass-produced automobile to Americans, he also gave them the first inexpensive, mass-produced tractor. While the Fordson tractor was not considered an example of superior design or engineering, it was the machine that quickly dominated the market when it was introduced. Its greatest virtue was its low price.

From the fateful day in 1917 when that first Fordson F, designed for Luther Burbank, rolled off the assembly line, every other builder had to redesign its machines to be competitive with the Fordson. Henry Ford, aided by the postwar depression, would put all but the most able tractor builders out of business by building a good machine cheaper than anyone else.

People have said that Henry was ahead of his time. He was always tinkering, taking things apart and putting them back together. His boyhood hobby was watchmaking, and one of his first jobs was operating a small milling machine. He spent his spare time reading magazines on mechanics. So when he happened to see a stationary gasoline-powered engine in operation in 1891, he quickly realized that it might be the power source he needed. He spent years experimenting, building a vehicle that could be powered by a gas engine.

The famous "Tin Lizzie" Model T came out in 1903 and sold for $850. In order to mass produce his gasoline-powered automobile, Henry had formed a corporation with a small group of investors. When the improved Model T made its debut, the American public was ready to buy automobiles and Henry Ford became a millionaire.

But Henry soon found himself at odds with his board of directors. He had put together the experimental Automobile Plow in 1907, and was thinking about mass producing a small traction engine. The directors of the Ford Motor Company were not interested. After a few years of argument, he formed a new company without any stockholders. This firm, which existed just to build tractors, was called Henry Ford and Son. Having revolutionized the automobile industry, Ford would now change American agriculture.

The tractor industry had known for years that Henry Ford was thinking of mass producing a tractor, and there was widespread speculation and discussion in all the major farm magazines. The general expectation was that Henry would build his tractors just like he built his cars—cheap and plentiful.

A lengthy article printed in the widely read and influential *The Country Gentleman* magazine in 1916 recounted Ford's comments and plans. Henry had visited the big annual tractor show in Fremont, Nebraska, and other attendees had been anxiously waiting the debut of the new Ford tractor. There was some expectation that Henry might bring it to the Fremont show, but he only talked about it in general terms with a news reporter. It would be called the Fordson, after his only son, Edsel Ford.

The first Fordson tractor, a twenty-horsepower, four-cylinder machine with a "unitized body," made its debut the following year. It was a somewhat primitive machine with every pound of excess iron discarded by the designers. They rejected the heavy frames used by steam tractor builders in favor of a one-piece casting—the "unitized body"—used to support the engine and attach the rear wheels. This long casting, which covered the drive shaft, also served as the frame.

The Fordson had several problems. One flaw was that the operator's seat was mounted over the end of the drive shaft, which meant that the seat would become warm and uncomfortable after prolonged use. Another flaw of the early Fordson was that it had no brakes. There was also a dangerous mechanical problem: the Fordson would sometimes dig in its wheels and flip over backwards, seriously injuring the driver.

Finally—and this was more of a problem to the marketing department—there were no plows or other implements designed to sell with the Fordson. Nearly every other tractor builder was a full-line agricultural machinery dealer. Farmers who bought Fordsons had to make do with whatever plows or planters they already owned. Henry, of course, did not see this as a problem. He didn't want to be in the implement business. Henry built and distributed autos—now he built tractors, too.

After a decade, technological advances and the marketplace would overwhelm the Fordson. There were better tractors, and other folks were building them. The Fordson would be viewed as useful only for dragging a plow around the field, and even that activity was sometimes hazardous.

Other builders were producing tractors that could be used for all the farm chores: planting, cultivating, and harvesting. The development of the PTO by International Harvester meant that the tractor could now also operate binders and mowers. Implements were being designed to match the tractors, and Fordson would soon be left in the dust, dependent on having other suppliers build suitable implements.

## 1918: The Fageol

After fifteen years, the tractor was out of its infancy and becoming an energetic, unpredictable adolescent. One of the most interesting tractors to make a brief appearance in the World War I tractor marketplace was the Fageol. Most tractor builders were designing machines for grain farmers or row-crop farmers. None of the conventional tractors were really suitable for orchard use. The

Fageol was built as an orchard tractor and could have been a tremendous success in its market. Yet that was not its fate. The low price of the Fordson combined with a serious post–World War I economic depression would force the Fageol Tractor Company out of business.

California's most important contribution to the tractor industry may be the unique machines developed especially for West Coast agriculture, machines like the crawlers, the Caterpillars, and the Tracklayers. The Fageol tractor also evolved

from a tracked model first designed for use in California orchards and vineyards.

Work in vineyards and orchards requires a special type of machine. Fruit trees and grapes are sometimes located on hillsides in very soft soil, with the root systems very close to the surface. Farmers need to work between the narrow rows and underneath overhanging branches without damaging the fragile limbs. The ideal orchard or vineyard tractor should be extremely stable and

have excellent traction. In typical orchard tractor models, the wheels are shielded by protective fenders and the driver's seat is placed very low, to prevent snagging on overhanging limbs. The early Fageol designs took all of these elements into consideration.

The Fageol tractor was engineered by an Oakland, California, company founded by two brothers from Iowa. Frank and William Fageol (pronounced FAD-jil) already had notable success building innovative motor trucks and pioneering designs for "parlor cars," as early tour buses were called. Their Fageol lumber truck business would eventually become the Peterbilt Company, maker of over-the-road trucks. Their Fageol bus factory would move to Ohio and become the Twin Coach Company, still an important builder of school buses. Fageol engines were powerful and reliable: the "Victory" engines used in their machines would be used to power military airplanes in World War I. With this expertise, it seemed quite reasonable that they wanted to open a division to build tractors.

The Fageol tractor was an unforgettable machine. The rear wheels were deeply toothed for maximum traction. The tractor was steered with a tiller rather than a wheel. And there were a series of four louvers decorating the engine cover, a signature that the Fageols also used on their other vehicles. It was altogether a stunning tractor, but it was doomed by the timing of its entry into the market.

The 1918 Fageol weighed thirty-six hundred pounds (1,634kg) and was difficult to maneuver. The new Fordson, introduced about the same time, was more agile and cost a third as much. In addition, the postwar depression would bring tractor production to a halt by 1921. Unable to survive the pressure, the Fageol was manufactured only from 1918 to 1922. Their San Jose, California, factory was closed and few tractors remain in existence.

These machines are prized by collectors for their interesting engineering and their scarcity; a few models can be seen in California.

# 1918: Waterloo Boy Model N

The John Deere Company was very late in adding a tractor to their catalogue. The company had been in the plow business for more than sixty years, building a wide variety of plows, harrows, and cultivators. Around the turn of the century, they began to expand their business by buying other companies that made farm equipment. They acquired a line of planters when they bought the Van Brunt Manufacturing Company, and wagons when they bought the Fort Smith Wagon Company. Binders, harvesters, and a buggy line were also acquired, the products modified to fit the needs of John Deere customers.

Gas tractors appeared in the John Deere catalogues and branch houses before 1910; some local dealers were selling the "Big Four" Model 30 built by the Gas Traction Company of Minneapolis. The John Deere Company was also exporting the Twin City Model 40 for sales in Uruguay and Argentina. Pressured by its own sales force and the competition, the John Deere Company began looking around for a tractor of its own. They turned to Joseph Dain, Sr., their former vice-president and a member of their board, to find a suitable tractor for the John Deere catalogue. He was asked to build a tractor that would sell for around $700.

Dain began immediately, finally producing an all-wheel drive model in 1915 that could be produced to sell for $1,200. The president of the John Deere Company, William Butterworth, balked. He felt that investing the capital to build such an expensive machine was unwise. The board debated the project, then put it on hold. Joseph Dain passed away.

The situation was at an impasse until another board executive mentioned that the Waterloo Gasoline Traction Engine Company was for sale. Buying a complete company was consistent with previous John Deere practices, so they did not hesitate. The board approved the purchase in 1918, and they were finally in the tractor business. The Waterloo Boy Model N, built between 1917 and 1924 and distributed by the John Deere Company, is considered the first John Deere tractor. It was also the first tractor ever evaluated in the Nebraska Tractor Test laboratories. The Waterloo Boy Model N was one of the smaller tractors of the time, with a two-cylinder engine, and rated as a 12-25 machine. It was sold to the small farmer and was not a particularly innovative design. In retrospect, its chief virtue seems to be that it put the John Deere Company in the tractor business. The Waterloo Boy name would be dropped around 1924 when the John Deere name first appeared on their new Model D tractors.

However, the name Waterloo Boy still has tremendous appeal to collectors and enthusiasts, and any of the rare early machines are avidly sought after.

**A farmer accustomed to steam tractors could feel at home on the big, broad platform of a Waterloo Boy. Compared to the competition of its day, the Waterloo Boy was a conservative, sturdy, durable design with a proven engine and drive train. It was bought by John Deere and became that company's very first tractor.**

# Weathering the Storms

*(1920 to 1936)*

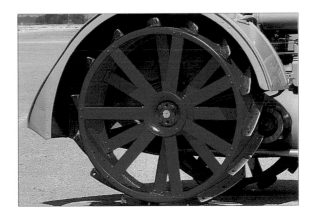

Power farming was a radically new idea, another marketing concept originated by the folks at Hart-Parr. It spread rapidly, picked up by agricultural colleges, county agents, and the agricultural public. Power farming was more than just extra horsepower, more than just owning a tractor—it became a system of work, a way to operate a farm with much greater efficiency and flexibility.

The progressive farmer of the nineteenth century used "scientific" methods; that's what was said to be the key to successful and profitable farm management. These scientific methods included soil analysis, crop rotation, matching the crop to the soil and climate, the correct applications of fertilizers, and the correct plowing and planting techniques. Suddenly the advent of power farming put an additional factor into the equation, boosting potential productivity to new levels. First scientific farming, then power farming. Now, just as American farmers were becoming intrigued by the benefits of power farming, the Fordson arrived to make it possible.

## New Companies, New Models

We generally think of the World War I era as a time of a booming American economy, but the war in Europe had brought turmoil and economic depression, and American agriculture was suffering. So was the tractor business. Histories say there was a time in 1920 when tractor manufacturing was completely suspended. Statistics from the *Yearbook of the Department of Agriculture, 1922*, tell us the grim news. The total number of gas tractors sold in the United States in 1920 was 162,988. The following year the sales were only 10,403. Having pushed his factories to build tractors to support Great Britain in the war, Henry Ford, along with other builders, found himself with an enormous surplus of tractors.

Ford was having difficulty selling his large inventory of Fordsons. First he shipped thousands of them to Europe to help reconstruction efforts after the war. Then he started dropping the price,

taking the loss in order to keep his factory in production. The price cutting put the competition into a tailspin. The price of a Fordson tractor dropped from $785 to $620, then again to $395. Only International Harvester had the market strength and deep pockets to match these prices.

International Harvester had another asset, the new McCormick-Deering Farmall, which was introduced in 1924 and is considered the tractor that finally pushed the Fordson from the field. In 1928, Henry Ford suddenly quit American tractor production altogether.

The 1920s were a time of enormous change in the tractor industry, and Henry Ford's price cuts were only a part of the equation. The number of tractor builders was continually rising. It seemed the wide acceptance of the automobile and the increase in the number of automakers led some car manufacturers to consider building tractors as well as autos as their market share was being squeezed. In 1920, *Farm Machinery and Equipment* magazine listed 166 tractor companies; in 1929 there would be only 47 of these left. Today only a handful remain.

In spite of the postwar depression, more auto companies and tractor companies were being organized and financed, and scores of different machines were put into production. In addition to the sudden surge of production from new companies, there were many creations from barnyard inventors who put together a machine or two from spare parts.

Competition was fierce, and every builder was looking at what the other fellow was designing. At the same time, developments and new designs in the auto industry were having an impact on tractors. The struggle for a share of the farmer's pocketbook was pretty fierce, too. The second half of the 1920s saw a tremendous struggle for control of sales territories, with numerous mergers and buyouts of agricultural implement dealers.

By the end of the decade, Hart-Parr would merge with Oliver and two others. Minneapolis-Moline would be formed from three smaller companies. The J.I. Case Company would absorb Emerson-Brantingham, and the Advance-Rumely

Company would merge with the Allis-Chalmers Company. Folded into these large corporations were dozens of smaller manufacturers of plows, planters, threshers, wagons, and rakes—all types of agricultural equipment.

By 1930 International Harvester controlled more than fifty percent of the tractor market, followed by John Deere with about twenty percent. The serious players remaining were J.I. Case, Oliver Farm Equipment, Minneapolis-Moline, Massey-Harris, and Allis-Chalmers. Every manufacturer had merged to survive.

## 1920: Gray

While the industry giants were battling for supremacy, there were still a few small tractor builders who managed to step aside to keep from being trampled. The Gray Tractor Company of Minneapolis specialized in orchard tractors, an industry that was generally overlooked. Fageol and small crawlers were being built for orchard work in the West, but there were few builders who were considering the need for an orchard tractor on the East Coast. C.H. Wendel shows pictures of the Gray tractor in his *Encyclopedia of American Farm Tractors*, but only a handful of Gray machines now exist in collections.

In an industry dominated by giants like Fordson, Case, John Deere, and others, a few small tractor builders survived by building very specialized machines tailored to regional needs. Manufactured in Minneapolis, Minnesota, the Gray tractor was built specifically for farmers with orchards, farmers who needed a different machine than those built for wheat or corn or row crops.

The prototype Gray machines were developed in 1908, but it was nearly six years before the machines reached the market. According to Wendel, the Gray remained in production until 1933, despite its relatively small market. This orchard model had a low profile and a low center of gravity. Similar to designs of other orchard tractors, the driver sat well back, but unlike other orchard tractors that only have protective fenders and a cowl covering the wheels and steering, this particular model had a protective shell covering the entire machine from front to back, just like a turtle. It was an exceptional machine for its time.

# 1924: Farmall

The McCormick-Deering Farmall tractor, built by International Harvester, probably remains the most revolutionary tractor ever built. Certainly it was the tractor that put Henry Ford out of production. It is still the favorite machine of folks who use tractors, prized for its utility as well as its extraordinary design and engineering.

The Farmall tractor was the favored child from an extraordinary gene pool, the merger of the powerful American Harvester Company with three competitors. The American Harvester Company itself was a partnership, formed in 1891 by competing harvesting equipment builders Cyrus McCormick and the William Deering Company. Negotiating with more than a dozen small companies for more than a decade, the McCormick-Deering partnership finally acquired three other competitors and the corporate name International Harvester Company was selected in July 1902. The Farmall tractor proudly carried that name and was known as the McCormick-Deering Farmall.

Adaptation and merger has been part of the International Harvester tradition since the beginning; it's their corporate way of life. When the company was formed in 1902, two of the five agricultural machinery builders were trying out gasoline tractors. The McCormick Harvesting Machinery Company and the William Deering Company had already been experimenting with gasoline tractors with some success. The McCormick tractor was known as the Mogul; the Deering machine was named the Titan. The new International Harvester Company sold both machines through their respective dealer networks.

By 1910, International Harvester had overtaken Hart-Parr as the nation's leading tractor producer. Of course, this was not difficult to do when only a few thousand tractors a year were being built. But International Harvester had a reputation as an industry leader to maintain, and they were determined to preserve their position. They wanted a tractor in their catalogue, and they wanted it to be a successful product.

The three Graces of Agriculture—plowing, harvesting, and threshing—each had its own corporate leader. While J.I. Case made his fortune building threshers and John Deere made plows, the largest builder of reapers, harvesters, and mowing equipment was Cyrus Hall McCormick. And like the John Deere and J.I. Case companies, the corporate founder was an inventive developer with enormous talent and drive.

The McCormick Company had been founded in 1840. Cyrus McCormick had personally promoted his machinery in many rural exhibitions, defended his patents in court, and promoted his machines at scores of county fairs. By the time he died in 1884, his organization was the largest company of its kind, an American legend. And like Case and Deere, McCormick also had a family and a corporation that were willing and able to carry on his legacy.

When they merged in 1891, both the McCormick and Deering companies had been in the agricultural machinery business for years and both understood the needs of their large customer bases. Both companies had developed steam traction engines for their large networks of branch houses.

At first, the new International Harvester Company hung on to the two tractors from the parent companies: the Mogul tractor and the Titan. When the small Bull tractor proved to be so popular, they quickly built smaller versions of both these machines. The very reliable Mogul 8-14 was crowned The King of Small-Farm Tractors. The two-cylinder Titan 10-20 was also extremely popular. With the success of these models, it looked as if a project to build a new tractor would be forgotten, for it would be difficult to scrap either of the successful older machines.

But everyone knew that Henry Ford would soon try to use his successful sales formula to build and market a tractor as well. So I-H was hard at work on a competitive model, producing the four-cylinder 8-16 in 1917. This tractor was produced until 1922, and would provide early competition for the Fordson.

The International Harvester Company was responsible for developing and introducing probably the single most important tractor in agricultural history, with the name that describes it perfectly, the Farmall. Designed by McCormick Work's Experimental Division under superintendent Bert Benjamin, it was the first machine that really could do it all: plow, plant, cultivate, and harvest.

Farmers and designers had talked for years about a "general purpose" tractor, a machine that could plant and plow as well as cultivate and harvest. This was the machine that delivered on all promises. It was small enough, simple enough, and cheap enough to be acquired, operated, and maintained by the average farmer.

Previous models from International Harvester had introduced the power take-off (PTO). The new Farmall improved this wonderful feature and added a few more besides. It rearranged the front wheels, putting two small wheels together in front and redistributing the weight of the tractor. It was the first machine to have the distinction of being

The famous Farmall Regular was the first true "general purpose" tractor, a machine that could plow, cultivate, mow, and power a thresher or sprayer. It didn't, as some people claim, introduce the power take-off (PTO), but it did perfect it. The compact little Regular can turn in only eight feet (204m)— marvelous agility then and now—a huge advantage for row-crop farmers. This one belongs to Denis Van de Maele, whose family has been buying and using Farmalls for as long as they've been made.

Ira Matheney's huge 1919 I-H Titan was a commercial success in the post–World War I tractor-buying frenzy, a powerful and durable first-generation design. But it is very big and bulky for its 10 drawbar/20 pulley horsepower—and it takes a real titan to crank it over until it starts, too!

what we now call a "rowcrop tractor," and it was able to turn in only eight feet (2.4m). This machine also introduced the use of frame mounting points to attach implements.

The Farmall went to the field without fanfare, but the machine sold itself and sales quickly exceeded expectations. It changed the shape of tractors forever, introducing the format of two small, single front wheels with two larger ones to the rear. The Farmall was the first tractor that farmers genuinely loved and would not trade. If all-around utility and owner affection is any indication of importance, the Farmall was the first machine that became a member of the family.

Field acceptance was immediately gratifying. This truly was an "all-purpose" machine that did everything. The experts had said that building an all-purpose tractor was impossible. But the Farmall Regular as it is known, soon became the farmers' favorite tractor. By the end of the decade, it was outselling its nearest competitor, John Deere.

In 1985, International Harvester merged again, this time with the J.I. Case Company, forming a new conglomerate known as Case International with the logo Case-IH. Today this group builds industrial equipment of all types—the tractor line is only one of its many products.

## 1928: Deere GP

Early tractors were being used for two types of
farm chores: they could pull an implement such as
a plow, or they could be belted to another piece of
farm equipment such as a grinder or an ensilage
cutter, providing the power to chop livestock feed.
There was a lot of farm work in between, however,
that tractors were not quite suited for, and inventors were working on adapting a tractor to fill
the gap.

One of the most time-consuming farm chores
is cultivating, which requires moving carefully
between the rows of seedlings, removing weeds,
and spraying for pests or applying fertilizer. A
motorized vehicle would make these tasks much
easier, but a suitable tractor had not yet been
developed. A cultivator tractor needed to be very
maneuverable in order to remove unwanted weeds
and leave the crop undamaged. Most builders had
been working on a type of tractor that was known
as a motor cultivator, a very lightweight tractor
with an arrangement of fixed cultivator teeth
attached between the rear wheels. Several builders,

including Henry Ford, had a working prototype
of a motor cultivator by 1910.

The John Deere General Purpose (GP) tractor
was the earliest machine offered by Deere that
could be used as a cultivator. A small cultivating
tractor known as the Model C had made its debut
in 1926, but due to some mechanical problems
only a few were made. The design was reworked
and appeared two years later as the Deere GP, or
General Purpose, tractor. The appearance of the
Farmall spurred all the competing builders to
add a cultivating tractor to their product line.
It looked like the engineers at International
Harvester finally had the technical details on
cultivators all worked out and others were quick
to imitate.

The General Purpose tractor was designed in
direct response to the popularity and flexibility of
the Farmall. The General Purpose was a popular
model for John Deere, selling more than thirty
thousand tractors in its eight years of production.
This John Deere was designed specifically for
cultivating chores. The front axle was arched to

Denis Van de Maele's 1934
Farmall F-30 is a trim, quick,
multipurpose tractor, part of the
I-H stable that helped transform
tractor design in the 1920s
and '30s.

provide clearance over row crops. The rear axle was lengthened to straddle three rows, compared to the Farmall's two-row reach. And there was an additional critical feature: a power lift system to raise and lower the implements.

The GP is now attractive to tractor collectors because of its size and its milestone technology. Various versions of the GP were offered during the eight years it was built, and owners refer to them by the "alpha designators" of their model. So the Deere GP-Orchard, called GP-O, is a General Purpose model fitted with special fenders for orchard use. In addition, there is a GP-Wide Tread (WT) and a GP-Potato (P).

## 1929: Case L

The Case Model L tractor appeared in the show-rooms late in the decade, but it represented a design effort at the J.I. Case Company that started in 1925. It was a giant leap forward in the way tractors were planned and built, perhaps the earliest tractor design shaped by the marketplace.

Until this time, the prevailing philosophy was that the engineer was the expert when it came to tractor design. The head of engineering or manufacturing at the factory had the first and last word, designing a tractor within the plant's manufacturing capabilities. He'd give the customer what he thought the customer ought to have. But as more and more farmers bought and drove tractors, they began to have some opinions of their own. So smart engineers began to listen to the customer.

The tractor builders at J.I. Case were watching the sales figures. By the early 1920s, most builders who had tried a crossmotor-type engine design had abandoned it in favor of a more conventional in-line engine. Case crossmotor machines may have been extremely solid and reliable, but they looked old-fashioned. Appearance was important because the farmer was becoming a selective buyer, looking for modern technology.

Customers seemed to like the Fordson's unitized frame. The Case Model L became the first tractor from Case built to use this sort of frame. Case improved their air filter, offering an oil-bath

**Model Ds roll off the end of the John Deere assembly line in Moline, Illinois, sometime in the 1920s. The shape of the D would change, as would production methods, but the basic machine and its big cast-iron engine would stay in production until 1953.**

Your tractor dollar is twice as big today

as it was 10 years ago . . . .

CASE Model 20-40 1921

1. Sold for $2500.
2. Two speeds - 2¹/₅ and 3¹/₅ miles per hour.
3. Power at only two points - belt pulley and drawbar.
4. The 20-40 was as good a tractor as money could buy 10 years ago.

The New CASE Model "L"    1931

1. Costs approximately one-half as much    ·    ·
2. Does nearly twice as much work per day    ·
3. Three speeds - 2¹/₂, 3¹/₄ and 4 miles per hour
4. Power at 3 points - belt pulley, drawbar and power take-off    ·    ·    ·    ·    ·
5. Does many more jobs much better and quicker
6. Costs much less to operate and to maintain
7. More profitable to own in every way    ·    ·
8. Twice as much work at half the cost    ·    ·

The New Case tractors, in all models, for all purposes, possess what tractor users have long wanted - dependable power, light weight, fast speeds, economical operation, easy handling and attractive appearance.

Case users consequently enjoy outstanding performance, lower production costs, more satisfaction from farming, and more profit - all at a price one-half of what it was 10 years ago.

J. I. CASE CO.
ESTABLISHED 1842  RACINE, WISCONSIN, U. S. A. INCORPORATED

**For Case and the other tractor makers, 1931 was a poor year. The ad was true, tremendous changes had occurred in just a single decade. Farmers were impressed and broke.**

influence was indirect. When the Case design team put together the Case L prototypes, they sent the tractors to the field to be used by actual farmers in actual field conditions. Until this time, any testing was generally done in a field behind the factory in Racine. A few tractors would be built and tested, and if the machine held together, the model would be put into production.

This process changed, however, with the advent of true field-testing and the input of marketing departments. The Case L was sent to Texas to be used in the fields. The design was then modified and refined. So the Model L not only represented a big improvement in engineering, it was a watershed machine because the customer's experience and wishes became part of the product.

It was a terrific tractor for the price. The Case Model L offered a forty-horsepower tractor for $792. A Farmall of the same approximate size was only a twenty-seven-horsepower machine and cost $875. And the thirty-seven-horsepower John Deere model was listed at $1,115.

## The Depression Years— Merger and Consolidation in the Industry

By the beginning of the Great Depression, more than half of the farms in the United States consisted of less than one hundred acres (40ha) of land. While small farms accounted for only about fifteen percent of land under cultivation, the tremendous number of small farmers indicated a consumer market with enormous potential. Clearly, a smaller, economical, general purpose tractor would be very attractive to a great number of buyers. Most small farmers were still using horses and mules, but as petroleum distribution became more widespread, rural America began considering gasoline tractors.

Tractors had changed dramatically during the 1920s due to increased competition within the industry. Automobile technology had a big

air filter that would keep dust and chaff out of the carburetor. Case built and tested a very solid transmission, a feature that would keep the Model L alive in the marketplace for nearly twenty years. Options included electric lights and an electric starter.

Farmers and salesmen had a lot to say about the design of the new Case L, even though their impact, too, as better carburetors, headlights, electrical starters, and all sorts of other improvements began showing up on tractors. Clearly, the highlights of the 1920s in the tractor industry were the introduction of the Farmall rowcrop machine and the introduction of a general purpose farm tractor.

The end of the decade saw many mergers, buyouts, and corporate failures in the tractor industry. An overheated market and overproduction led to speculation in the tractor business, just one more contributing factor in the stock market crash of 1929.

By 1931, the Aultman and Taylor Machinery Company of Mansfield, Ohio, and the Advance-Rumely Thresher Company of LaPorte, Indiana, had become part of the Allis-Chalmers Corporation. Like International Harvester, Allis could now show at least ten long-established agricultural equipment builders in their family album.

The J.I. Case Threshing Machine Company had purchased the Grand Detour Plow Works and the Emerson-Brantingham Company and had acquired the remains of the J.I. Case Plow Works, thereby ending the confusion in the marketplace regarding the similarly named companies. When Case absorbed the Emerson-Brantingham Company it also got its tremendous line of agricultural equipment and tractors. This

acquisition brought Case not only the E-B product lines but the Heider and Rock Island tractors, too, which were carried in the same catalogue. Case folded the assets from all these companies into its own product lines, keeping the best products and discarding the others. The number of manufacturers listed in *Farm Machinery and Equipment* annual magazine decreased by more than two-thirds in less than a decade. The economic conditions of the depression thirties would force even more changes.

## 1930: The Field Marshall

While the Great Depression slowed demand for American tractors, it provided a small window of economic opportunity for the development of British tractors. One of the most interesting tractors ever built was a remarkable single-cylinder diesel tractor from the Britannia Works of Marshall in the town of Gainsborough. The prototype for the Field Marshall was first exhibited at the 1930 World Tractor Trials at Wallingford.

The tractor world had never seen a machine

A Case Model L with a small combine in tow, harvesting small grain in the early 1930s. This combination represented modern power farming at the time. The combine has its own gasoline engine for power, and the Model L easily replaces the dozen or so horses that would have previously been used for a combine this size (six for the morning's work, six for the afternoon's). Now two men are performing the work done by a dozen on a threshing crew, and they are doing it in style and comfort.

quite like this one. It had no magneto or spark plugs to start the motor but was started by lighting ignition papers in a holder protruding from the combustion chamber. The radiator, sitting in a transverse position, was also uncommon. The early Marshall tractors were fitted with a fuel pump and injector of the company's own design and manufacture.

The earliest model was rated as a 15-30. Power was provided by a single-cylinder engine that had an eight-inch (20.3cm) bore by ten and a half-inch (26.7cm) stroke and burned fuel oil. For the first several years of production, each machine was custom made to suit the individual needs of the

customer; no two tractors were exactly alike, either in engineering or color.

By 1933, the design was becoming a little more standardized and the tractor had undergone noticeable design changes. The Marshall Works were also building an 18-30 model, and the company stopped using their own pumps in favor of a more reliable Bosch fuel pump and injector. But even though design and engineering became standardized, the customer still had several options on color. It is estimated that fewer than one hundred 18-30s were built.

Then, in 1934, the company came out with a small machine, a 12-20. This solid little diesel tractor with rubber tires proved to be a popular model, with an estimated five hundred units sold, many of them exported to Australia, New Zealand, and elsewhere.

Not surprisingly, the 3.5 ton (3,178kg) Marshall, priced at £315 (approximately $1,575 at the time), had some competition: the English-built Fordson, which was available for only £155 ($775). The Fordson was lighter, cheaper, and had enough power to drive a thresher; the only advantage the Marshall could offer was the use of diesel fuel. But the machine had its loyalists since Marshall was well known in England as a purveyor of quality, especially when compared with Fordson.

At the end of World War II, the Marshall company divided its product line into agricultural tractors, the Field Marshalls, and road construction tractors, the Road Marshalls. With various modifications and improvements, the tractors continued to be produced until the mid-1950s, when high-speed diesel engines from Fordson International and other American tractor builders became available at considerably lower cost.

## 1930: Case C and CC

The J.I. Case Company entered the depression years with the Model C tractor, an economy-sized version of their popular Model L. While the choice of the letter "C" designator seems somewhat illogical in the alphabetic scheme of things, it actually designates the model chosen from a series of prototypes for a Fordson-size tractor. Case wanted to build a little tractor they called a "general purpose" machine, one that would combine the best features of the Fordson and the Farmall.

In the late twenties, Case was already working hard on adapting the Model L as a cultivator, and the engineers were having difficulty redesigning the spacing on the wheels without losing maneuverability. It was not possible to simply build a smaller version of the Model L, and the optimal solution forced the Case engineers to completely redesign the front end.

The final product was a pair of small tractors, one, a cultivator with a gooseneck front end to provide crop clearance, the other, with a more conventional front end similar to the larger Model L. The Case Model C and the Model CC (cultivator) were both rated at 17-27 horsepower, about half the size of the Model L, which came in at 27-40. Both had four-cylinder engines. While the big L weighed well over five thousand pounds (2,270kg), the smaller C was just over four thousand (1,816kg). Although the large L was the machine that appealed to Case's traditional customer base, it was the smaller C that was designed to compete with tractors in the Fordson class.

The Model C and the Model CC were like fraternal twins: they were born at the same time with plenty of similarities but yet were ultimately different. Both were very popular machines with an eleven-year production run. The Model C sold well even through the depth of the Depression: more than 20,000 total units were moved. The Model CC did even better, with 28,652 sold. And, like the John Deere tractors, the C series tractors

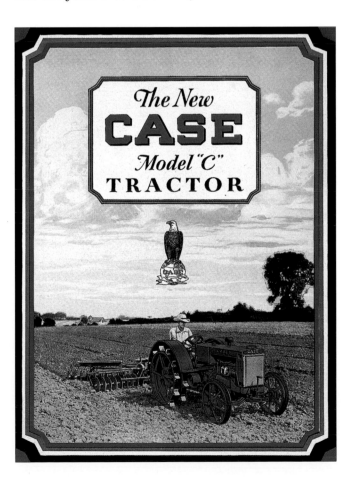

The Model C was the Case entry into the "general purpose" tractor market and it was a success despite the Depression, staying in production for over eleven years with more than twenty thousand units sold.

were offered in models with alpha designators to puzzle the uninitiated. There was a C-Industrial (CI) and a very rare C-High Clearance (CH)—only seven models were built. The C-Orchard (CO) was also rare, as was the CO-VS, a Vineyard Special.

## 1931: Farmall F-30, F-20, and F-12

The first McCormick-Deering Farmall rowcrop tractor was built by International Harvester. It later came to be called the Farmall Regular and was a big hit when it was first introduced in 1924. By 1931, it had been in the marketplace for half a dozen years, enough time for copycats to build similar machines and enough time for farmers to use the machine over several seasons and register their opinions. Rowcrop tractors were on the farm to stay.

The Depression was slowing sales for everyone, including tractor makers, and International Harvester really had only one model of tractor to sell. Clearly, they would need to offer something more if they wanted more customers. Their answer would be a series of updated rowcrop machines. The Farmall F-30 was first produced in 1931. It was longer, stronger, and heavier than the earlier Farmall rowcrop tractor. The 1924 Farmall had been rated as a small machine, able to pull two plows. The new F-30 was larger, rated for three plows. It was a competitive match in size and strength to the Model L from Case. The first Farmall F-30 was a three-wheeled rowcrop tractor, but later models were available with four wheels.

Produced from 1931 until 1939, the F-30 was joined in 1932 by two smaller machines: the F-20, basically an updated version of the first 1924 Farmall, and then the F-12, the smallest of the three. Small, medium, and large—International Harvester had a model in each size to get them through the Depression.

The F-20 differed little from the older Farmall Regular. The designers juiced up the engine a little, improved the steering a bit, and added a four-speed transmission. It was enormously popular all through the Depression years, selling 150,000 models by 1939.

The F-12, built for smaller budgets, had features that were similar, but it was a much smaller tractor than the F-20. It was first offered with a single front wheel, and it was a good small tractor for the "one-horse farm." It offered a seven-foot (2m) turning radius and was rated to pull one plow. With this outstanding line, I-H cornered more than fifty percent of the market.

RIGHT: Ira Matheney's 1937 Farmall F-20 spent most of its life working on a farm near Olympia, Washington, before Ira adopted it. Sixty years after it was built, the F-20 is nearly all original, except for the paint and decals.

OPPOSITE: Farmall's popularity was due in great measure to its versatility. Here is Denis Van de Maele's machine, equipped with its cultivator, one of many useful implements designed just for this tractor.

# 1934: Allis-Chalmers WC

In the annals of tractor history, the Allis-Chalmers Company is credited with two pioneering innovations. They were the first tractor builder to offer machines with pneumatic rubber tires, a tremendous improvement in both the performance of the machine and in comfort and safety for the operator, and they were the first to market their tractor with a standard package of implements at a fixed cost. This pricing strategy would help them maintain their precarious foothold in the marketplace during the Great Depression.

The 1934 Allis-Chalmers was put through the Nebraska Tests twice: once on inflated rubber tires, once on traditional steel rims. Allis management had collaborated with Harvey Firestone to

put a tractor on real rubber tires, and the result was a milestone machine for the early 1930s. In another unusual innovation, a real stroke of marketing genius in the dismal days of the Depression, the Allis-Chalmers management decided to change the color and the shape of their machines. The tractors were painted an eye-popping Persian orange and the shape of the sheet metal cowling was smoothed and contoured, offering the first attempt at streamlining ever seen on a tractor.

In 1935, two more Allis models were released in two versions: one with pneumatic rubber tires and one with traditional steel rims. Both the Allis Model U and the Model UC, a rowcrop version, went through dual tests in Nebraska at the test labs. Pneumatic tires offered such a dramatic improvement in the way a tractor operated and handled that all the other tractor builders quickly began to offer them, too. Until this time tractors were commonly offered with either steel wheels or steel wheels with hard rubber treads, used for warehouse vehicles and tractors driven on roadways. A few other tractor builders had experimented with the inflatable tire in the early 1930s, and one enterprising Florida dealer even put a few tractors on big oversized airplane tires for better traction in sand. It soon became clear, to skeptical users, that rubber tires provided numerous improvements over steel wheels.

A tractor with rubber tires was more maneuverable, offered better fuel efficiency, and allowed the farmer to operate at higher speeds. The first rubber tires were sold on an Allis in 1932, but the 1934 Allis WC was the first machine actually put through the Nebraska tractor tests on pneumatic tires. By the end of the decade, rubber tires had become standard and an estimated ninety percent of all tractors sold had them. There was also a tremendous market for refitting older tractors with new pneumatic rubber tires.

Rubber tires kept Allis alive during the Depression years, but so did their sales strategy. Allis offered package pricing, the first manufacturer to do so. Walt Buescher tells the story of super-salesman Harry Merritt and how Harry

kept Allis in business during the depression in his book *Plow Peddler*.

According to Buescher, Harry Merritt was a force to be reckoned with. He is described as "a giant, a man and a half tall, a man and a half wide and two men heavy." Harry Merritt did not carry calling cards because he felt that if a man didn't make enough of an impression to be remembered, a calling card wouldn't help. Harry was a memorable salesman.

Harry put together an irresistible package to increase tractor sales. He offered a tractor with a single-blade plow, a one-row cultivator, a five-foot disc harrow, and a little forty-inch pull-type combine as a package for "only $1,093.25." He knew that farmers would like to bargain him down to an even $1000. Tractor sales shot up.

Merritt also found a way to get farmers to spend an additional $150 on rubber tires for their tractors, even in the middle of the Depression. He had his sales manager plow the infield at state fairs using a rubber-wheeled tractor. Then they hired Barney Oldfield and Ab Jenkins, two of the world's most famous race car drivers, to race each other around a state fair racetrack on a pair of Allis tractors with rubber tires. A reported record of 67.4 miles per hour (108.5kph) was set. Walt Buescher recalls that although the racing had nothing to do with farming, it put Allis tractors and rubber tires in front of a lot of customers. And farmers would talk about the race, providing a great deal of inexpensive public relations for the company.

The introduction of rubber tires combined with the Allis-Chalmers color scheme, styling, and marketing to put the company ahead in the tractor race. Allis became the third largest tractor producer in the middle of the Depression, behind International Harvester and Deere. This success, however, lasted only a short time. Oliver was ready to introduce the Oliver 70, a move that would turn the tractor world upside down. Much later, in 1985, the Allis-Chalmers Company was folded into a conglomerate known as Deutz-Allis; today after other buyouts the company is known as AGCO.

OPPOSITE: The Allis-Chalmers G was another interesting model, introduced around 1948, and designed to compete with the "little" machines introduced by John Deere and other builders. It is popular with collectors today because of its small size. This little tractor, restored by Carmin Adams, features a four-cylinder engine and produces about ten horsepower.

# 1934: Deere D, Then GP, and Back to A

There is nothing more confusing to a newcomer in the tractor world than having to relearn the alphabet according to tractors. The model designations seem to make no sense and a Model B rarely follows a Model A. That's just the way it is in the tractor world, where each manufacturer had its own reason for tagging a particular model. So with that caveat in mind, we will look at a few of the models built by John Deere during the 1930s.

The John Deere Company began to hit their stride in the tractor business just before the Depression. The designers at John Deere had been badly frightened by Henry Ford and his tractor back in 1917. When the Fordson first took to the field, John Deere was selling the Waterloo Boy, a machine they would soon discard in favor of their own design.

As John Deere looked over the competition, the trend seemed to indicate the need for a smaller, lighter, more maneuverable tractor, one that looked and handled more like an automobile. John Deere engineers settled on a design they called the Model D, the first two-cylinder tractor with the John Deere name. It turned out to be a tremendous success, with a thirty-year production run starting in 1924.

But no sooner did they produce their Model D than the Farmall appeared and another competitive challenge had to be met. Deere's answer was a little rowcrop that could be used as a cultivator, which they called a Model C. The Model C machines were only built for one year, 1928, and were rebuilt and renamed the next year.

But the short-lived Model C introduced a device that used the power from the engine to raise and lower the cultivator. It was nearly as important an innovation as the power take-off. This innovation was quickly adapted (or copied) by other builders. The new improved Deere model was designated "GP" for General Purpose and it would include all the improvements of the Model C plus a more powerful engine.

The General Purpose lived up to its name, and the basic machine was outfitted with all sorts of useful options and modifications. Depending on which set of options the farmer ordered, a new GP could be designated a GP-WT (Wide Tread) or a GP-P (Potato). The potato tractor was a wide-tread machine with a narrower rear axle for use in potato fields. Smaller than the Deere Model D, the General Purpose tractor would be in the catalogue until the mid-1930s.

The Model D would be joined by the Model A in April 1934 and would become the most popular tractor in John Deere's history. This affordable tractor was flexible and could be used for both plowing and seeding as well as cultivation. It was medium-sized and capable of pulling three plows. The Model A goes down in tractor history as the first tractor series to apply hydraulics to both power and lift implements.

The basic A was very adaptable and came with a wide range of styles and options, a veritable alphabet soup of possibilities. The Model A with a single front wheel was called an "AN," or A (narrow); the A with an adjustable wide axle was an "AW" or A (wide). The letter "H" was added to designate "high-clearance." Deere enthusiasts refer to their pets by their initials, an "ANH" or an "AOS," for example. The latter is a Model A with orchard (O) attachments, streamlined (S). They each look a little different.

Before the development of the power-lift, a farmer had to mechanically raise and lower the implements at the end of each row—an arduous job. With power-lifts, the implements could be raised simply by pulling a lever. A hydraulic device, powered from the engine of the tractor, provided the muscle. This not only saved the arms and backs of the farmers, it was also a tremendous time-saver.

The popularity of the Model A paved the way for a smaller model the next year. The Model B, designed as a back-up tractor for larger farms, could pull a single plow and was especially suited for small farms. It came with the same alpha designators as the Model A. And like the Deere

Farmall rowcrop and the John Deere General Purpose tractors indicated that Case would have to develop a real rowcrop tractor of their own to stay competitive. They had tried calling the Case Model C a cultivator, but International Harvester's Farmall still cornered most of the market.

Introducing a new model gave Case the opportunity to bring out an "economy-sized"

Dale Hartley's old 1935 Case RC is a good example of the many light, inexpensive, and capable rowcrop tractors that became so popular toward the end of the Depression, in the latter half of the 1930s. This one was a good investment; it still starts easily and is ready to work the lush, rolling fields near Hamilton, Missouri, where it spent its career.

Model A, the letter "H" was added to designate "high-clearance." This pair of Deeres was highly successful and greatly loved. There were more than 300,000 units built of each model over their long production runs.

## 1935: Case RC

Could anyone afford to bring out a new product in 1935, the depth of the Great Depression? Along with everything else in the economy, tractor sales were down, but Case felt that there was no choice. The continuing popularity of the

machine, something smaller than the L and the C models. The Case RC came out in 1935, suited especially to the small farmer. It could pull one plow and was a good machine for the rowcrop farmer, who was now making up more than sixty percent of the sales market. The Case RC sold well enough, but a dramatic event was about to take place that would once again turn the tractor industry upside down: the introduction of the stunning Oliver 70.

# Chapter 4

# Improving the Hybrids— The Introduction of Styled Tractors

*(1936 to 1939)*

In the depth of the Depression, just when the economy seemed to be at its darkest, a glorious new tractor was introduced that caused farmers to look up and predict a bright new day. This was perhaps the prettiest tractor ever built: the beautiful Oliver 70. More than just another pretty face, the new Oliver dazzled everyone with an assortment of features that had never been seen before. Most important was the introduction of aerodynamic styling.

This feature made the introduction of this tractor a watershed event.

## Styled Tractors

It was named the Oliver 70 because it ran on the same seventy octane petrol that powered most cars. For the first time, a standard tractor really ran on ordinary gasoline. It had a six-cylinder engine while the John Deere had a two-cylinder and most other tractors were four-cylinder machines. It had an electric starter, factory headlights, and rubber tires. And it had remarkable styling: it was the first tractor with a swept-back, aerodynamic look, making it the nicest-looking tractor anybody had ever seen. As an added bonus it had a factory-built AM radio! In 1935 even most automobiles did not have a radio. You could practically hear the mad scramble as engineers at International Harvester, Case, and John Deere ran for their drawing boards.

The major tractor builders were quick to react, hiring preeminent designers experienced in industrial styling. John Deere hired Henry Dreyfuss of New York City, known for his streamlined design of the Super G Constellation

for the Lockheed Aircraft Company. International Harvester signed up Raymond Loewy, noted for his streamlined locomotives for the Pennsylvania Railroad. Improving the aerodynamics of a tractor that moved around the field at a blazing three miles per hour (4.8kph) would not have much impact on performance, but tractors had clearly moved into the modern age, and sleek appearance was important.

## Oliver 70

The milestone Oliver Hart-Parr 70 is a unique tribute to the historic Hart-Parr Company. On the nose of this beautiful machine is the name of the maker, Oliver, and below in smaller letters is the name of the tractor industry founder, Hart-Parr. These two companies had merged in 1929, along with Nichols and Shepard and the American Seeding Machine Company. The four had joined to become the Oliver Farm Equipment Company. The Model 70 was the last tractor to carry the names of both companies and the first to put the Oliver name on top and in large letters. Good-bye forever Hart and Parr.

The new Oliver 70 lived up to its heritage. Combining streamlined beauty with the performance of a high-compression engine, it was an outstanding machine for 1935. The early field tests were a little embarrassing for the competition. The I-H F-20 looked clumsy and the Case C looked feeble in comparison. One story says that the John Deere machines kept running out of fuel. Suddenly every other tractor looked old-fashioned and unwieldy.

The Oliver 70 also came with a unique new implement system attached to the tractor with mounting tubes that fit through the frame. It was easy to attach cultivating tools, plows, or any other implement to this new Oliver. This "Easy-On, Easy-Off" rig combined with the power-lift made the Oliver 70 an extremely useful and versatile tool. The genius behind the design of the Oliver 70 and the integrated implement system was provided by a pair of brothers from the Oliver company in South Bend, Indiana: Herman and Rudolph Altgelt.

## Hart-Parr and Oliver

It's rare for a tractor company to be mentioned in general history books for contributions to technology. Hart-Parr is one of these rarities. The extraordinary Hart-Parr Company proudly carries the title Founders of the Tractor Industry, figuratively and literally—this proud message is cast right into the radiators of the Hart-Parr tractors. The Hart-Parr Company gave the world the gasoline-powered tractor in 1902, founding an industry and helping the average farmer to become an independent businessman.

Some thirty years later, in 1935 to be exact, the new Hart-Parr Oliver corporate partnership would make tractor history a second time. They introduced a new tractor with a streamlined shape and an engine that ran on regular gasoline. This was the beginning of the modern tractor, a machine with everything, including a six-cylinder engine and a radio. Hart-Parr Oliver deserves the label as the most historic tractor company in America.

Today the company has dissolved, melted into one corporate merger after another. The first merger occurred in 1960. Acquired by the White Motor Corporation of Cleveland, a builder of large trucks, the Oliver tractor division was then merged with two others in the late sixties, Cockshutt Farm Equipment of Canada and Minneapolis-Moline. The product of this merger was subsequently renamed the White Farm Equipment Company.

Then the White Farm Equipment Company was in turn acquired by Texas Investment

Corporation, which then became part of Allied Products Corporation. Allied already owned the New Idea Farm Equipment Company so they merged the two farm implement lines into the White–New Idea Farm Equipment Company. The old Hart-Parr plant in Charles City continued building tractors under the New Idea name until the building was demolished in the summer of 1995.

## Restyling the Oliver Line

Although the new Oliver 70 had been introduced just two years earlier, the entire line of Oliver tractors was reconsidered in 1937. The 1935 Oliver had been offered in four configurations: a rowcrop, a standard tread, an orchard model, and a high-crop model. By 1937, the name Hart-Parr was dropped from the logo of all machines. The styling was further refined on all models. The orchard version came with sleek, wide aerodynamic fenders. There was also now a tough little industrial model for work on loading docks and in factories.

**OPPOSITE: One of the prettiest examples of the breed, the Oliver 70 was restyled in 1937 to this handsome conformance and set a standard for all other manufacturers that lasted for years.**
**BELOW: Here's the original Oliver 70, a pioneering design even before being restyled. This one is a 1935 Row Crop example. Like the rest, it has a six-cylinder high-compression engine designed for seventy-octane gasoline, a power plant legendary for its smooth running and excellent power. That engine was a first for tractors, as was its electric starter and lights, plus an innovative system for attaching implements. The Oliver 70 broke the mold.**

By 1937, there were some other changes and refinements in the Oliver product line. Although the stylish Oliver 70 was a high-compression engine operating on regular gasoline, many farmers were still using kerosene (distillate) as a fuel. The new Olivers came with a choice of engines.

There were some changes in model sizes, too. Way back in 1930, the newly merged company had brought out a large tractor capable of pulling three to five plows. It was rated at 28-44 and was carried in the catalogue with a new name, the Oliver 90. Now it, too, would be overhauled and dressed in a stylish new suit.

Finally, to fill the gap between the Oliver 70, which was a twenty-four-horsepower machine, and the big Oliver 90 at forty-four horsepower, a new, midsize Oliver was offered. The Oliver 80 made its appearance in 1938 in the same range of configurations: gasoline or kerosene, a four-wheel regular or a three-wheel rowcrop. Looking over the entire Oliver lineup, it was a staggering variety of machinery for a smaller company to offer during the Depression years.

But there was still more to come. A tiny tractor, the Oliver 60, would be added as something of an afterthought in 1941. It was added to the lineup to contend with a new and formidable competitor, the 9N from Henry Ford. Ford came back into the tractor game in 1939 with a small machine featuring the Ferguson hitch system. The senior tractor builders at Oliver remembered the Fordson, and they knew that they should pay attention to Henry and his tractors. By the end of the decade, all the major manufacturers were taking up the Ford 9N challenge and bringing out smaller models.

## 1936: Farmall Sees Red

When International Harvester built the Farmall, they created one of the most popular tractors of all time, successfully countering John Deere and prevailing over Fordson. But the impact of the gorgeous Oliver 70 was a development that surprised everyone. The quickest tactical solution that International Harvester could implement was

these was the Farmall A, a styled machine with a radically new feature: the engine was offset to the left side while the operator's seat and steering wheel were offset to the right. This gave the driver an open view of the ground beneath the tractor to clearly see the crop. Leowy called the improved view "Culti-Vision," and this feature proved to be a very practical development. The Farmall A replaced the older F-12 and F-14.

The larger Farmall B, the second in the Loewy I-H series, replaced the older F-20. With its twenty-five-horsepower motor, this tractor was designed to compete with the John Deere B. All of the Farmalls designed by Loewy had nearly identical frames so the implements and attachments could be used interchangeably, no matter which size Farmall the farmer bought.

Finally, in 1939, the International Harvester Company brought out the largest of the Farmall series, known as the Mighty M. It, too, replaced a popular earlier model, the F-30. With big horsepower and five forward speeds, it was the favorite machine for large-acreage farmers. This tractor was eventually offered in 1941 with an optional diesel engine. Although there was a substantial increase in price for the diesel, it proved justifiable since the tractor quickly returned the initial investment in economical operating costs.

**ABOVE:** Everett Jensen's Orchard version of the Oliver 70. Orchard tractors are rare and fragile; the extensive sheet metal tends to rust away or get discarded, so this example is particularly precious. This is probably the most attractive version of the most stylish basic tractor design of the 1930s, and one of just a handful known to exist.

**RIGHT:** The Farmall M was introduced in 1939, just before World War II, and quickly became one of the most popular tractors ever used on American farms. Heavy and strong enough to handle a three-plow hitch, this machine proved to be a real workhorse until the 1950s. It remains a popular machine today.

to repaint their tractor line, changing from years of drab, dull gray to a fiery red-orange. Then they quickly called for reinforcements.

The designer that International Harvester brought in to help their product line was Raymond Loewy, an individual who was noted for the beautiful design of the Studebaker automobile. Now Loewy turned his attention to I-H, and no detail was overlooked. He redesigned the corporate logo for International Harvester, changed the appearance of their dealerships, and went to work on the product line.

Loewy's first job was to clean up the tractor grille and introduce streamlining. The first I-H tractor to carry the new Loewy look was the International Harvester crawler, known as the TD-18. With a contoured metal hood and newly curved front end, it was outstanding in its bright red paint.

Shortly thereafter, a new line of tractors appeared with a stylish new motif. The first of

# 1937: Allis Baby B

Harry C. Merritt was an Allis-Chalmers vice president and one of the most innovative managers the company ever hired. He introduced rubber tires to the tractor industry and forced the competition to change sales techniques when he introduced a specially priced package of a tractor and four implements for less than $1,000. He was also engine and adjustable tread width, it was a small machine that was strong enough to pull a plow and adaptable enough for all farm chores. It was so cute that buyers just wanted to put it in their pockets and take it right home.

Priced below $500 in 1937, this would prove to be a very popular little tractor. It sold more than 127,000 units during its twenty years of production. Its primary competition was the John

the inspiration behind the Baby B, the tiny Allis tractor that sold for less than $500. Now every farm, no matter how small, could own a tractor.

It would be one of the smallest machines built that could still be considered a real tractor. It weighed a petite twenty-six hundred pounds (1,180kg), a fluffball when compared with the machines of a decade earlier that weighed five or six tons (4,540kg or 5,448kg). With a four-cylinder

Deere Model L, a small machine weighing around a ton (908kg) that came out in 1937. But the Deere L had a two-cylinder engine using regular gasoline as fuel. It put out less than ten horsepower compared with the fourteen horsepower of the Allis. Overall, the better machine for the small farmer who could only afford one tractor was the little Allis.

## John Deere Gets Stylish— Introducing Models G, H, L, and GM

The John Deere Company also went shopping for a renowned industrial engineer to develop a design that would compete with the Oliver and the Farmall. They engaged the services of Henry Dreyfuss, designer of a wide variety of modern products. Originally a designer of sets for Broadway shows, Dreyfuss also worked for the Bell Telephone Laboratories and the Lockheed Aircraft Company. Dreyfuss quickly evaluated the current Deere Model A and Model B machines as well as the little Model L. Within a month of his arrival, some prototypes were ready to show to the management. The short-term solution was to quickly revamp the existing models; the long-term solution was a completely new machine.

Deere engineers brought out three styled machines in 1938. With a new hood and radiator covers, the styled tractors proved to be both functional and beautiful. The styling program at John Deere was to continue for the next eleven years, until every option on every model received the Dreyfuss touch.

Just as Henry Dreyfuss was being contacted, John Deere was already introducing a new line of tractors, the Model G. Production had started in 1937 on the Model G, a large machine capable of pulling three large plows with thirty-five horse-power. The Model G would escape Dreyfuss styling until 1942, offering a somewhat old-fashioned-looking alternative to the conservative farmer who might distrust the fancy new styled tractors. The Model G was designed for farmers with large acreage, who needed a heavy tractor but also wanted some of the capability of a row-crop tractor. The Model G had a long production run: it was built until 1953.

The John Deere H, which had the same power as the older Model B, was a general purpose, row-crop machine in the mama-bear size. Designed for the small farmer who used only one or two plows, the Model H came out in 1939. Like all the other John Deere tractors, it, too, could be ordered with a variety of optional configurations to meet the specific needs of each field situation.

For example, the Deere HWH was a model variation that appeared in 1940. The plain vanilla Model H was a rowcrop tractor with dual narrow tires on the front. The second letter, W, says that it came with an adjustable wide front axle to provide a crop clearance of twenty-one inches (53.3cm). The HWH came into being when a grower requested front wheels that could be adjusted for height as well as width. He wanted to be able to run his tractor above double-wide tomato beds rather than in the furrow.

The adjustable front wheels were not originally a John Deere option, but since the customer wanted them, the local dealer found a way to install them. He used some extensions in the front axle that were manufactured by the John Deere Cultivator Works, and the grower was set up to run above the tomato beds. The HWH proved to be such a popular machine with Western growers that by 1941 it was advertised as the tractor of choice for bedded crops.

Deere brought out one more model near the end of the Depression. Sometimes considered the baby of the John Deere line, the Model L, referred to as the "Little" or the "Light," had about nine horsepower. Introduced in 1937, it was first designed with slightly offset steering to give the farmer a good view of the crop. It was probably most suitable to the buyer who had a large yard rather than a small farm. International Harvester picked up—and even improved—on this idea almost immediately, introducing its offset Farmall tractor with the trademarked "Culti-Vision" system in just two years. The little L, an important transitional tractor for Deere, was eventually offered in a styled version.

The LA showed up in 1940, a slightly heftier version of the little L, with an added four horsepower. The back wheels were also a little wider. This popular machine continued in production until 1947, when the entire product line was replaced by all new Deere engineering, starting with the Model M.

# 1938: The Minneapolis-Moline U-DLX

It made perfect sense in Minneapolis weather, and it seemed logical on the drawing board. But this tractor just didn't sell. The Model U-DLX (deluxe), offered by Minneapolis-Moline designers in the late thirties, was something created by marketing theory rather than by marketplace reality. It was a tractor that looked like a car, with windshield wipers, a heater, and a radio. It was known as the "Comfortractor," and it proved to be too far ahead of its time.

In theory it was a very practical machine for bad weather. Any number of cold-weather chores, such as dragging hay out to livestock in the snow, would be a lot easier in a tractor with an enclosed cab and a heater. And for Depression-era farmers with limited resources, forced to choose between buying a tractor and an auto, this machine could serve both functions.

While the U-DLX was certainly attractive, it was simply not practical for most farm chores. An important factor in selecting a tractor is ease of mounting and dismounting. Another is visibility, the ability to look at the furrows in front and quickly check the performance of the implement

**They call it "Prairie Gold," the unique color and style of the great Minneapolis-Moline tractors. This one is a 1951 Model U from Jerry Silva's stable of beefy blondes.**

in the rear. The U-DLX cab configuration limited the operator's access and visibility all the way around. Priced at more than $1,800, reportedly only 125 machines were produced. Because of their interesting design and their rarity, and because many contemporary tractors now look like the "Comfortractor," the U-DLX tractors are highly prized by collectors. It was a machine that was fifty years ahead of its time.

## 1939: Case Gets Hot— The Flambeau Series

After the Oliver 70 introduced styling to the world, it became clear that all other tractor builders would have to find designers who could "pretty up" their machines and provide some of the streamlined details that farmers were beginning to expect on their tractors.

The entire concept of styling was a little foreign to tractor builders. Streamlining was for airplanes, race cars, roadsters, that new line of diesel locomotives—all machines that epitomized high speed. But for a tractor—a machine that rolled around the back forty at a top speed of three miles an hour (4.8kph)— styling seemed simply ridiculous.

The Case company, still one of the biggest agricultural machinery producers, was also caught with a dull and boring product line. The sales numbers proved it. The three Depression-era

tractors in the product line were all painted gray, so Case management decided that the quick and easy solution was a new coat of paint. They picked a highly distinctive color that would really stand out in contrast to John Deere green and other colors on the market. Case decided on a screaming hot red-orange. They called Flambeau Red. It was a smart move.

The Flambeau series of Case machines would remain on the market all during the World War II and Korean War years, selling until 1955. Of course, a change in color was just the beginning; the demands for styling necessitated important engineering updates and refinements. While all of the older models from Case—the heavy L series, the long-running C series models, and the R series that was first introduced in 1935—received Flambeau Red paint schemes, it was obvious that all models would need to be replaced by machines that could live up to their flashy exterior.

Change was just around the corner. The first newly styled, Flambeau-to-the-floor tractor from Case was the Model D, introduced in 1939. At the time, the best-selling tractor in the industry was the Farmall H, the very versatile two-plow tractor. Looking at what the market wanted, the Case design department also developed the Case Model S, a scaled down and very stylish version of their new Model D. With completely new engineering and styling, the S offered a new engine and a new transmission.

The Case Model S was introduced in 1941 along with its companion rowcrop tractor, the SC. The popular rowcrop SC sold about six times as many units as the S, which amounted to about forty-seven thousand units during the war years. It was finally replaced by a newer model in 1951.

The little tractor could also be ordered as an SI (Industrial) or SO (Orchard). Advertising brochures frequently showed the SC being driven by younger members of a farm family, perhaps to demonstrate that it was suitable for use by Junior (or Mom) while Dad was away at war or involved elsewhere in a civic role.

This new size of tractor finally allowed Case to match the competition at Deere and International Harvester model for model. Besides the medium-sized D, rated as a three-plow tractor, and the smaller model S, which was capable of pulling two plows, there was also the big Model LA, a five-plow machine, and the petite Model V series, smaller than the Model S. The new styling and the new engineering of the Flambeau line would put Case tractors firmly in third place in the tractor world.

A fresh Case SC rolls off the assembly line about 1940. Competition was stiff and all the surviving manufacturers labored mightily to come up with the flashiest, most stylish and technologically advanced designs. The SC's orange-red paint (christened Flambeau Red) is visible at three miles (4.8km) and was one of the selling points for the machine.

# Chapter 5

# Harvesting the Fields—Hitches, Hydraulics, and Horsepower

*(1939 to 1960)*

PREVIOUS PAGES: (Left) International Harvester's bright red tractors dominated much of the post–World War II marketplace, for several reasons. One was that they were built for the ages, and many are still serving honorably. Another was that they were designed to do just about anything, with a five-speed transmission, reliable and powerful engine, and at economical cost. But that bright red really captured the imagination of the farmer, and stole his heart. This one stole the heart of Donald Henderson; it is his beloved W-4, a model built from 1940 to 1952 in many variants.

(Right): Tractor styling got a little goofy in the 1950s, but the styling for just about everything—not just tractors—was the same way. Ford added this little medallion to the bowsprit of their Golden Jubilee tractor, a special model issued to celebrate the company's fifty years in the tractor business. OPPOSITE: International Harvester's beautiful, showy Super M was a tremendous success in the postwar marketplace, with plenty of power, a superb engine, and excellent agility (for a tractor). This 1952 example belongs to Robbie Soults, and like his other old iron, it runs like new. Many similar versions, but with faded paint, still serve on farms around the world.

In the long run, the tractor industry benefited from the World War II years. Raw materials were scarce: steel and rubber went to build ships, planes, and other armament rather than tractors. But engineering innovations from wartime weapon designs, improvements in hydraulic systems, diesel engines, and new production ideas would eventually make their way from the builders of tanks and airplanes into the auto and agricultural machinery industries.

Designers waited until the war was over to introduce new tractor models. The restraint was partly due to wartime shortages of materials and partly due to patriotism and good public relations.

The war would make some permanent changes in tractor design. Chrome trim disappeared from machines such as those from the Case company, never to return. Ford came out with a wartime model, the 2N, which had no electrical system and steel wheels due to material restrictions. But the biggest changes after the war would come from new ideas. Tractors would have improved hydraulics to power the implements, new implements with easily operated hitches, and more horsepower.

## The Handshake Agreement and the Lawsuit

There are a few episodes in tractor development that have been recognized by historians as watershed events. The Handshake Agreement that brought the Ford-Ferguson tractor into existence is one. Although the war years of the early 1940s brought constraints to the tractor industry, those years also brought demand for increased food production. There was a tremendous need to produce enough food to feed not only the fighting armies but to replace crops destroyed in the war zones. A small, inexpensive tractor with a universal hitching system would be just the tool to assist in the war effort. The Ford tractor with the Ferguson system was introduced in 1939, and it quickly proved to be the ideal machine to meet this need.

It was an unbeatable combination, a cheap tractor with a universal attachment system for implements. Hydraulic lifts allowed the implements to be raised and lowered, attached and detached in a minute. It could easily be operated by one person. The Ford 9N became a very popular machine very quickly. Like the Model T Ford car, its main virtues were its mechanical simplicity and its price. It was a terrific tractor, the product of a singular partnership. It was produced by a unique contract, a handshake between Harry Ferguson and Henry Ford.

The moment was historic for a number of reasons. The invention of the three-point hitch itself was a tremendous benefit to the farmer. Combining the hitch with the most affordable tractor in the world assured that the small, independent farmer would benefit. And sealing a contractual agreement of this magnitude with just a handshake, with nothing but the good faith and good word of the participants, was a tremendous expression of respect and trust. It's no wonder that the Handshake Agreement was considered an historic event.

## Ford Meets Ferguson

It was the inventive Irishman Harry Ferguson who brought Henry Ford back into the tractor business. Ferguson had designed and perfected a unique hitching system for plows and other implements, and he needed a tractor builder as a partner. Ferguson and Ford first met around 1917, and the two teamed up to install and demonstrate a Ferguson plow system on a Model T automobile. Ford then offered Ferguson a job but Ferguson declined, even when the salary offer was raised. He wanted to be independent. Even though Ferguson had developed a useful plow and hitch, his primary interest was in racing cars and automobile design, not tractors. So Ferguson went back to Great Britain and back to his drawing board.

Harry Ferguson was born on a farm in Ireland. Like Henry Ford, he disliked farmwork. He, too, showed an early interest in anything

mechanical and spent a great deal of time as a lad in his brother's auto repair shop. In 1909, at the age of twenty-five, Harry Ferguson had designed, built, and flown the first airplane of any kind in Ireland. Airplanes, motorcycles, racing cars, tractors—Ferguson was interested in all sorts of mechanics. A lightweight tractor plow was only one of his many patented inventions.

The Ferguson designs for implement hitches were obvious and brilliant, simple solutions that anyone could look at and say, "Of course, why didn't I think of that?" But as others point out, the devil was in the details. Other designers made so-called universal hitching systems for attaching

into the soil, bringing the nose of the tractor up and even flipping the tractor over backwards. It took a very quick reaction from the farmer to depress the clutch and disengage the gears to prevent damage to the plow, the tractor, and the driver. Accidents resulting from encounters with

implements to tractors, but only Harry Ferguson engineered a design that was perfect.

## Harry Ferguson and the Duplex Hitch

The design for a new hitching system originated in Ireland, a place with rocky soil that was a terrible hazard to tractors pulling a plow. The point of the plow would snag on a rock with very dangerous results. Unless the tractor could be stopped immediately and the plow freed quickly, the wheels of the moving tractor would keep digging

snags were not uncommon, often resulting in dismemberment and sometimes death.

The solution was to design an implement hitch that had some play, so that the hitch itself would ride up and down, allowing the plow to roll over rocks and other hidden obstacles. Harry Ferguson designed and demonstrated his Duplex Hitch before 1920, but while it solved the problems of snagging and dangerous bucking, it created another problem. The plow now rode up and down in the soil, making the depth of the furrows uneven. Farmers need to plant seed at a uniform depth, so some sort of system was required to ensure that the plow point always returned to a specific level. Keeping furrows level and even was known as draft control, and before tractors it was traditionally a skill the plowman practiced by eye and by hand. Ferguson patented his Duplex Hitch, as well as another device called a Floating Skid, which he designed to help keep the plow at a constant depth.

## Hydraulics and the Three-Point Hitch

Harry Ferguson, an inventor, not a manufacturer, looked around the United Kingdom for a tractor maker as a partner to help mass produce his hitch system. He first made arrangements with British builders and even toyed with building his own tractor. The most common tractor around England was still the Fordson: about 800,000 of them had been imported or manufactured for use by British farmers since World War I. Although the 1929 crash and the Depression curtailed tractor production, Ferguson was still interested in some sort of joint venture with a tractor builder, and Henry Ford was an obvious candidate.

In the meantime, Harry Ferguson and his design team continued to refine their hitch system. One of their most important develop-

ments was the installation of a system of integral hydraulics that was powered by the engine of the tractor. It allowed plows and other implements to be raised and lowered with a hydraulic lift that the farmer could operate from the seat of the tractor.

By the 1920s, self-contained hydraulic units consisting of a pump, controls, and motor were developed and in general use, paving the way for applications in all sorts of machinery from locomotives to airplanes. The hydraulic system adapted by Harry Ferguson used the power from the tractor motor itself to provide the power to raise and lower the implements.

In addition, Ferguson and his designers were adapting his Duplex hitching system and perfecting the draft control device on a hydraulic-powered system. The Ferguson team developed a simple yet ingenious refinement to improve draft control. They used an arrangement with three attachment points on the tractor and equipped them with ball joints for flexibility. Two attachment links were located under the tractor in front of the rear axle. The third point was up high, just behind the driver's seat. The three points formed a virtual triangle, making the entire system quite stable. The upper link was fitted with a sensor that kept the implement at a uniform depth.

A prototype of the hydraulic-powered hitch and lift system was demonstrated for Henry Ford, who was immediately interested in the invention. Henry wanted to be back in the tractor business, but he was still not interested in providing implements for his machinery. The Ferguson hitch was designed to be an integral part of the tractor; it could be used with any available implement. The two inventors, Henry Ford and Harry Ferguson, struck a deal.

This agreement, which the two men entered into in 1938, consisted of a discussion followed by a handshake. There was no written corporate contract, just a verbal gentlemen's agreement between two individuals. According to one source, the

agreement called for Ferguson to be responsible for the design and engineering of the tractor and hitch. Ford would manufacture the tractor and its integrated hitch system. Ferguson would then distribute and sell the tractor. Either party could terminate the agreement at any time for any reason.

Given the reported terms of the agreement and the personalities and irascible natures of both Ford and Ferguson, many observers were skeptical about the longevity of the contract from the beginning. Historian Randy Leffingwell humorously observes that the Handshake Agreement proved as good as the paper it was written on. In fact, the agreement lasted for six years.

## 9N—The Ford Tractor with the Ferguson System

The first 9N tractor, known as the Ford tractor with the Ferguson system, was introduced in June 1939. The Ford Model 9N was a tremendous hit because of the Ferguson system, and about ten thousand tractors were sold in the first year. When the Ford-Ferguson appeared in dealer showrooms in 1939, it offered an electric starter, a power take-off, and the three-point hitch that allowed eighteen different implements to be either attached or removed in about a minute, all for $585.

During the war years, tractor production became extremely restricted. The Ferguson-Ford continued to be built, but without essential components. A modified tractor, known as the wartime Ford 2N, was sold without a starter, a generator, or rubber tires. The model number commemorates the year it was introduced, 1942.

The Handshake Arrangement worked pretty well during the World War II years, although there were already forces churning that would make the contract unworkable. Henry Ford was elderly and ill. He lost his only son, Edsel Ford, in 1943. Nearly eighty years old, Henry suffered a series of strokes that left him seriously incapacitated. His twenty-eight-year-old grandson, Henry Ford II, took over the management of the company in 1945.

Young Henry and his new management team quickly reviewed the existing business arrangements and noticed that Ford was losing money by building tractors and giving them to Harry

Ferguson, Inc. to sell. Clearly this did not make good business sense. So in mid-1946, Henry Ford II informed Harry Ferguson that Ford was planning on ending the Handshake Agreement within one year. Then young Henry directed his engineers to design a new Ford model to replace the Ford-Ferguson 2N. This new tractor would also take its model number from the year of its appearance—8N for 1948.

The new model 8N would continue to feature an integrated Ferguson-type hitch system and a number of other improvements. It would be distributed through a new company, one owned by Ford. Ford would pay no royalties to Ferguson for use of the Ford-designed tractor and hitch system. Then the ailing Henry Ford senior passed away in 1947. Now left with no possibility of discussing the revised contract arrangements with his old friend, Harry Ferguson sued.

The litigation between Ford and Ferguson would hamstring tractor design after World War II. Although other tractor builders were all designing their own variations and adaptations of the three-point hitch, none dared release them into full production until the lawsuit was settled. The basic Ferguson patents for the three-point linkage and draft control were running out when the suit was filed in 1947. But the lawsuit itself was really two legal actions covering a complex list of complaints; patent infringement and royalties were only some of the issues.

The litigation entered the courts in 1948, and the Ferguson vs. Ford suit was finally settled in April 1952. Ferguson immediately took out full-page ads in all the newspapers to announce the outcome. Ferguson had sued for $251 million. According to the judgment, Ford would have to pay $9,250,000 to Ferguson. All the other tractor builders breathed a sigh of relief and looked at their own three-point hitch development. The Ferguson patents had now lapsed. Very soon every major tractor manufacturer would bring out their own variation of a hydraulic-powered universal hitch with draft control.

## Horsepower—Diesels for Tractors

The tractor industry usually seems to ignore developments in related businesses like the automobile industry and the industrial equipment industry. Then an advance in technology forces the tractor builders to stop and take notice. Streamlined styling caught the tractor industry by surprise in the mid-1930s. There was another extremely important technology that was also introduced in the 1930s that made its way into the agricultural world: diesel engines.

The Fordson 9N was a big hit when it was introduced in 1939 with its three-point hitch, but the no-frills model 2N produced during the war years was sold without headlights or even a starter. This beautifully restored Fordson is owned by the Kuckenbecker Tractor Company.

HARVESTING THE FIELDS (1939 TO 1960)

## Diesels Warm Up

Caterpillar came first, installing a diesel engine in its Model 60 as early as 1931. According to Randy Leffingwell in his book *Caterpillar*, Oakland, California, contractor Henry J. Kaiser installed Atlas diesel engines in Caterpillar crawlers to be used at a construction project on the Mississippi River in the 1920s. The diesel engines stressed the chassis on the Caterpillar 60, but the machines lasted long enough to get the job done.

Then in the late 1920s, a demonstration in Africa forced the Caterpillar company to get serious about diesel engines. A cotton-plowing contest was set up between a Caterpillar Model 60 and a British machine from G.J. Fowler with a Benz diesel engine. As Leffingwell reports, Caterpillar got licked! Afterward, the defeated engineers immediately ordered a Benz diesel from Germany and began redesigning the Model 60 to fit the new engine. Known as the D-9900, the diesel-powered prototype was up and running in 1931. Caterpillar introduced this model into the marketplace in 1932. Caterpillar began selling a diesel-powered crawler, and International Harvester was watching.

## Diesels at Full Throttle

The Depression took its toll on American businesses, and everyone's revenues were down. Railroads reduced operating expenses by replacing coal-burning steam engines with more efficient diesel machines. In the early 1930s, gas for tractors sold for around fifteen cents a gallon while diesel fuel averaged about five cents. Railroads became big consumers of diesel fuel as steam locomotives were replaced with economical diesel engines.

Diesel engines, developed by German inventor Rudolf Diesel, were introduced to American manufacturers about the same time that the Otto gasoline engine was gaining popularity. Rudolf Diesel was awarded a patent for his technology in 1893, and by 1897 he had signed contracts to build his engines in the United States. Diesel engines found immediate applications as power plants for ships and trains. Installations of diesel engines for various uses were widespread after 1900.

A Diesel engine was exhibited at the 1915 Panama-Pacific Exposition in San Francisco. Tractor builder C.L. Best saw it and was very intrigued. By the mid-1920s, he had installed a diesel engine in his crawlers, and by 1932 the Caterpillar company had a diesel-powered crawler in production.

The research and development departments of the big tractor builders were interested in the possibilities of using diesel engines. The J.I. Case Company experimented with a few diesel engines in the 1930s, including a Cummins diesel and a Hasselman diesel engine, designed and imported from Sweden. But interest in diesel engines for tractors was short-lived when the government decided to tax diesel fuel.

Thousands of diesel engines were used by railroads and ships, and in large commercial power plants. Looking at a way to increase their own revenues in the Depression years, thirty states promptly moved to tax diesel fuel in 1934. While the major railroads could absorb the increased expense, the small farmers could not. The expense of acquiring a newer, more expensive diesel tractor combined with the increase in fuel expense put diesel tractors out of consideration for them. For these reasons, diesel tractors, except for the earth-moving crawler-type equipment, did not become popular until the early 1950s.

Diesel engines began appearing again in agricultural tractors during the World War II years. International Harvester was the first to produce a diesel farm tractor in 1941, offering a diesel engine in their Farmall "Mighty" M. The price of the Farmall MD (Diesel) was a whopping fifty percent higher than a conventional Farmall M, but it quickly paid for itself in fuel savings. Installing diesels in their farm machines came as a result of their experiments with diesel engines in their crawler models.

International Harvester had already spent several years refining a diesel engine. The first International Harvester diesel tested in the

OPPOSITE: John Deere introduced their first diesel, the Model R, in 1948, and it quickly proved to be a very economical machine. In production for six years, it is of great interest to collectors since only about twenty thousand units were sold.

Nebraska laboratories had been the TracTractor, a crawler machine brought out in 1937 to compete with the diesel-powered crawlers built by Caterpillar. The TracTractor set fuel efficiency records and proved to be a very economical machine to operate. The TracTractor was a kind of engineering crossover, the crawler that forced conventional tractor builders to consider diesels.

It's hard to keep a secret in the tractor business, and the competition quickly noted the new development. Other manufacturers followed the leader and added a diesel engine to their catalogues as an option, but it was nearly fifteen years before diesel engines became commonplace on the farm.

John Deere brought out their diesel in 1949; the John Deere Model R was tested in the Nebraska Tractor Laboratory and performed well. Economic conditions were tight immediately following the war and expensive diesel tractors were not as popular as other models. But the early success of Farmall's diesel caused other builders to look again at offering a diesel. The J.I. Case Company tested their diesel, the Case Model 500,

in late 1953. Oliver also tested a diesel tractor that year, eventually becoming something of a leader in the tractor industry by offering a range of diesels in various sizes of tractors.

## Tractors—Bigger, Better, and Beyond

Styling, hydraulics, and new engines—it's hard to envision how tractors could get much better. By the 1950s, tractors came with reliable starting systems and a choice of fuel systems for the engines—regular gas, diesel, or LPG (liquefied petroleum gas). Each tractor builder had four or more sizes to offer, and each size frequently came in several models—regular, rowcrop, utility, and so on. In addition, each machine could usually be customized with a wide range of options to fit every field situation.

If anything, there were too many choices and too many tractor builders for the marketplace, and the result was inevitable. There would be another round of mergers and downsizing as tractor productivity improved. Fewer and fewer farms were needed to feed Americans. But there were still some tractor innovations that needed to be offered to farmers, and the smart designers were quick to identify them. Tractors needed to be safer, and tractors needed to be more comfortable. The tractors of the future would pay attention to both needs.

While the Ford tractors had a profound impact on the agricultural machinery business, it is amusing to note that Ford was, for many years, a one-model tractor builder: the 9N. Other makers were offering a wide range of sizes, speeds, and options, but while Henry Ford was alive, there was only one model of Ford tractor available. It's also important to remember that Ford was not an agricultural machinery builder but primarily an automaker. During the 1960s and 1970s, Ford expanded their tractor lines, adding new models

and implements. Their expanded product line included lawn and garden equipment as well as giant four-wheel-drive articulated tractors. In 1991, the Ford tractor organization became part of a conglomerate known as the New Holland Holding Company. Known commonly as New Holland, the group includes Ford, the New Holland Company, the Versatile Tractor Company of Canada, and Fiat Agri.

## Oliver

The Oliver Company was able to stick with four models during the World War II years, barely holding its own against the competition. The Oliver 70, 80, and 90 were developed before the revolutionary Ford 9N appeared, and although Oliver had a "motor-lift" for implements, they were still at a distinct disadvantage. The fourth Oliver model, the little Oliver 60, was introduced in 1940 specifically to compete with Ford. It had a four-cylinder engine, smaller than the other Oliver models, but it was available as a rowcrop, a standard tractor, and an industrial machine.

Oliver waited out the war with the models that were already in its stable. Like all the other major manufacturers, Oliver was ready with new designs and new engineering the minute the war was over. In 1947, they brought out all their newest engineering, and it was pretty impressive. They offered six-cylinder machines with a range of fuel options: gasoline, kerosene-distillate, LPG, and diesel. The Hydra-Lectric hydraulic lift system for implements was an option. The new Oliver Models 66, 77, 88, and 99 replaced the earlier machines.

Oliver realized that the competition from Ford was not going away. Even though Ferguson was taking Ford to court, the American farmer was still interested in owning a small utility tractor with a three-point hitch. Oliver engineers developed the Model 55, introducing it in 1954. Here was an Oliver tractor that looked like a Ford and handled like a Ford. It was a utility tractor that came with a three-point hitch and either a gasoline or diesel engine. And it was one of the earliest tractors made with a twelve-volt electrical system. Oliver was still in the game.

The design engineers in the Oliver research and development department had been quick to identify some features they felt ought to be on their future machines. Oliver saw operator comfort as a high priority, and they envisioned a tractor with an enclosed, weatherized cab with tinted windows, a heating and air-conditioning system, and a full range of instruments such as a tachometer on the dashboard. Oliver thought the farmer ought to have a comfortable adjustable seat, a radio, and cellular communication. A cigarette lighter was a desirable accessory, too.

There were a few other features that would make farming safer and more comfortable. Non-skid surfaces should be standard on all work platforms, better lighting on both front and back would improve safety by improving visibility, and the hitch should be shielded to prevent mishaps.

Oliver research and development engineers had a long wish list. Improved instrumentation, including a clock or timer that would show the hours of tractor use, was on the list. Another item was a simple button on the dashboard that would cause implements to be raised or lowered. And an automatic garage door opener would be a big help in opening gates and barn doors.

Few, if any, of these innovations would ever actually be offered on Oliver machines, although it's interesting to note that nearly all of these features are currently offered on contemporary tractors. Safety equipment, including a seat belt and a roll bar, was soon built into every tractor. This last important feature was soon mandated by the Office of Occupational Health and Safety (OSHA) and became known in the tractor industry as ROPS (Roll Over Protection System). But the Oliver tractors had long left the field by the time these features became standard.

## 1957: Oliver Super 44

Oliver had some terrific ideas, but they seemed to lose their engineering edge in the mid-1950s. Perhaps one of the first indicators of the loss of market focus was the introduction of the Oliver Super 44, a strange-looking little tractor that was a departure from conventional Oliver practice in many ways. It was built in Battle Creek, Michigan, rather than Charles City, Iowa.

Produced for only a couple of years, from 1957 to 1958, this was the first utility or all-purpose tractor built by Oliver that actually carried the name "Utility." Although Oliver had built the Oliver 55 in 1954 as its answer to the Ford 9N, this curious little machine is the one that is usually called a utility tractor.

Its design quickly became the target of numerous jokes. The Super 44 had some really good features, but the execution of the design detail was unconventional. The seat and steering wheel were offset, like the Farmall with "Culti-Vision" of earlier times, in order to give the driver a good view of the field. The frame was low and wide, making it a very stable little machine. And it offered an internal hydraulic system to operate a three-point hitch.

But all these progressive features were ignored when the purchasing public got a good look at the

This Oliver OC-4-31G belongs to John Boehm and poses for the camera in John's extensive walnut orchard. The OC-4 is a gasoline-powered little crawler perfectly suited for orchard duties where soft soil and sometimes tight rows can make navigation hazardous; it spent its working life in a vineyard where its very narrow profile was essential. It was built in 1957, and you could get one with either a three- or four-cylinder Hercules engine; this one has the four-banger G option.

steering column. Then the snickers and the pointing began. The Super 44 looked like it had been designed and assembled by a committee. The steering assembly protrudes from the grille in front and the tie-rods overhang the front of the frame like a bumper. It looks strange, as if the mechanical details had not been not carefully considered. Built for only two years, it is interesting to collectors because of its rarity.

After the Super 44, Oliver's days were numbered. In 1960, Oliver celebrated the twenty-fifth anniversary of the introduction of its six-cylinder engines. Within a few weeks, the company was gone, acquired by the White Motor Corporation of Cleveland.

## John Deere

Once again the tractors from Ford proved to be formidable competition. John Deere management began to realize that the plain little Ford 9N was changing the way that farmers thought about tractors. Deere's response was the Model M, a plain little tractor of their own. They called it a general purpose, utility tractor. While all tractor builders were itching to put their own version of the Ferguson system on a tractor, they wisely decided to wait until the lawsuit was settled. So although each major builder had a model that matched the Ford 9N in size, none dared offer a feature that exactly matched the Ferguson system. In 1947, John Deere introduced its Powr-Trol system, which allowed implements to be powered by hydraulics.

## 1949: John Deere M

The Model M, a little eighteen-horsepower tractor with an electric starter, came out in 1949. Offered in three very useful variations, the little tractor was equipped with a Touch-O-Matic hydraulic control system that raised and lowered the implements on each side independently. The John Deere Company offered twenty integral attachments for use with its little M. These were all important features, but the M still didn't quite match the Ford-Ferguson.

In addition, Deere quickly upgraded their older Model A and Model B tractors, adding an electric starter, lights, and high-compression engines. Finally they brought out a diesel tractor in 1949, calling it the Model R. Now Deere felt it could comfortably compete with both Ford and International Harvester.

**The first numbered models in the John Deere line appeared in 1952 and then quickly disappeared in 1956, so this Model 60 belonging to Doug Peltzer is pretty rare. It sports a two-cylinder engine but is beefed up to produce forty horsepower.**

# John Deere by the Numbers

Upgrades and new attachments were just a stop-gap measure. John Deere knew it would need an entirely new model line, and they prepared for a complete revamp by 1952. The alphabet designations for their models, which had served since the first John Deere D rolled over the horizon, were all swept away in favor of model numbers. Now the new model sequence made more sense. The Model 40 replaced the model M, the machine to match Ford. Starting with the small Model 40 utility tractor and working on up, the new models were: the 50, the 60, and the diesel 70. In 1955, the big sixty-seven-horsepower diesel Model 80 appeared.

Live hydraulics and a live power take-off system that were now independent of the main clutch were offered on all models. Deere tractors were offered with a choice of fuels: you could specify an engine that ran on gasoline, All-fuel (distillate, tractor fuel, or gasoline), or LPG (liquefied petroleum gas). Deere became the first manufacturer to offer factory-installed power steering on tractors in 1954.

Change is the norm in the tractor business, and no sooner did Deere have all options on all models than they decided to upgrade. Still competing with big International Harvester, Deere changed their model numbers and their paint scheme in 1956. The Model 40 became 420, the Model 50 became 520, and so on. The new model number indicated a number of changes. Deere added a new, comfortable seat that they called Float-Ride, and offered their entire line with a two-toned paint job: green and yellow. Five-speed transmissions, the Custom Powr-Trol (Deere's answer to the Ferguson system), a twelve-volt electrical system, and lengthened clutch and throttle levers were also offered. The competition was overwhelmed. There was, however, one thing a John Deere tractor still lacked: a four-cylinder engine.

## John Deere 420

The Green Machines were still competing with International Harvester on one side and Ford on the other. Ford celebrated their fiftieth anniversary in 1953 with the introduction of the Ford Jubilee model. John Deere's challenge to Ford had been the Model 40; now it was upgraded and designated the 420, equipped with even more options.

The Deere 420 offered a double handful of models and an increase from twenty-five horsepower to twenty-nine horsepower. Models included a general purpose tractor, a standard tread, orchard, utility general purpose, a two-row utility, a special (26-inch [66cm] clearance), a row-crop, a crawler, and a high-crop (32-inch [81.3cm] clearance). The options included a five-speed transmission, a continuously running PTO, and a three-point hitch. An LP gas option was also added. John Deere was serious about competing with Ford.

## The End of Johnny Popper

In 1960, an era ended at John Deere when the two-cylinder "Johnny Popper" was finally retired in favor of a machine with a four-cylinder engine. It was the end of the Long Green Line, which stretched back forty years, an eternity in the fast-moving agricultural machinery industry. The end of the Johnny Popper era marks an important milestone for tractor enthusiasts; machines built after 1960 just aren't as much fun.

Machines of the future at John Deere would include giant tractors of 215 horsepower with power steering, power brakes, and automatic transmissions. The first four-wheel-drive tractor had been introduced in 1959. It was clear to the John Deere management that tractors were going to get a lot larger and a lot heavier, and that the two-cylinder engine had reached the upper limits of its capability.

## The J.I. Case Company

The J.I. Case Company tractor lineup was in pretty good shape during the war years. Their best selling Flambeau Red tractors put them in third place among tractor builders, right behind International Harvester and John Deere. Although Case did not develop a new model specifically to compete with the Ford, their popular V and later VA models, first introduced in 1940, did meet some of the Ford competition by being small and affordable machines. The V model, built from 1940 to 1942, was phased out and replaced by the VA, which continued in production until 1955.

The Case Model LA was introduced in 1940 and was the largest model in the line at the time. This beautiful example belongs to John S. Black of Slater, Missouri.

The Case VA was designed to compete with the John Deere Model B. Both were small tractors rated at around twenty horsepower, capable of pulling two small plows. Both came in a wide variety of configurations. The Case VA could be ordered as a standard tread, rowcrop, industrial, orchard, high-clearance, or offset high-clearance tractor. There were military versions as well. Options included the Eagle Hitch, Case's answer to the Ferguson three-point hitch, which was introduced in 1949. Case had offered a motor lift for implements, but it would be replaced by hydraulic control in 1949.

Unlike Deere, International Harvester, and the others, Case did not revamp its product line just after World War II. Instead, Case made continuous upgrades and improvements to its four models, ranging from the small V/VA to the Model S, the Model D, and finally the large LA. The Case product line would not see a major change until the early 1950s. Then things would really change!

## The Case 300, 400, and 500

Daring, dazzling, dynamic, diesel—the advertising adjectives kept piling up when Case finally introduced the Case 300 as a replacement to the outdated S series tractors. The engineering on the Case 300 was outstanding. The tractor came with built-in hydraulics, two different transmissions, and four different fuel options: diesel, gas, LPG, or distillate. It was offered in six models, including general purpose, utility, rowcrop, industrial, orchard, and high-clearance versions. It could be ordered with a four-, eight-, or twelve-speed transmission. But one thing the Case 300 did not have was a pretty face. Even though this particular model offered a two-tone paint job and updated styling, its grille tried too hard to be modern and ended up looking dated.

Fortunately, the styling of the larger models was more conventional. The larger sizes, models 400 and 500, were equally dazzling in engineering features. They had a push-button starter and a system of "live" hydraulics that allowed the operator to operate the implements powered by the

hydraulic systems independently of the power train. They had power steering. The 500 had a six-cylinder diesel engine with fuel injection. By the late 1950s, the styling would change again, with squared-off noses and an aggressive thrust typical of contemporary auto design. But there were other corporate developments that would have a more profound impact on the future of Case tractors.

The J.I. Case Company tractors would change radically in the late 1950s as industry changes focused the Case organization on new markets. Only a small part of the J.I. Case Company business was tractors. They had started out in the agricultural machinery business as a builder of harvesters. Their entry into the steam traction industry had made them a world leader in road building and industrial equipment. Case had other product lines besides agricultural equipment.

In the mid-1950s, the Case company acquired the American Tractor Corporation, a builder of crawlers and a loader backhoe. Now the Case company expanded their industrial line, adding dozers, brush rakes, and fork lifts. Case did not completely neglect their agricultural business, revamping their tractors and changing both the styling and the model designations. Models VA, S, D, and LA were replaced with models 300, 400, and 500. These three sizes were later joined by models 200, 700, 800, and 900. The new models had a new paint job. Two-tone paint was fashionable for tractors, and the Case colors were now Flambeau Red and Desert Sand. And Case machinery offered all the options that the other builders sold: diesel engines, a choice of fuels, and automatic transmissions.

**OPPOSITE: The industrial Case SI was produced between 1942 and 1952, with a rugged nose designed for warehouse and loading dock use during the World War II years.**
**ABOVE: Now equipped with push-button starters and automatic transmission, this Case 400 lets the driver rest comfortably on a new hydraulic lift system and has improved power take-off as well.**

International Harvester offered this small tractor, for small "truck" farms and big gardens, and for lawn mowing and light cultivating, and like John Deere's LA, it was a big success. While the offset engine makes the tractor look like it has been in a bad wreck, the actual result is to give the operator unobstructed vision ahead and below, a tremendous advantage when cultivating young plants.

In 1960, Case held a tremendous sales and marketing tractor show in Phoenix, Arizona. The brochure headline said "12 Distinct Power Sizes…124 Models." The show featured an old-time tractor pull, with a Case tractor equipped with Case-o-Matic transmission besting a comparable John Deere. The expansions proved to be too much. After several years of financial difficulty, Case became part of the Tenneco conglomerate in 1967. In 1985, Tenneco also acquired International Harvester, and the two organizations were combined to form Case I-H.

## International Harvester

The Farmall models refined by designer Raymond Loewy just before the war carried International Harvester until the mid-1940s. The famous "Culti-Vision," with the offset steering wheel, offered loyal customers all the innovation they needed for several years. But by 1947, it was apparent to the agricultural machinery manufacturers that major industry adjustments were going to be made. Raw materials for new tractor models became available again, but the war had forced American farmers to operate more efficiently. Fewer farms were feeding more Americans, so fewer tractors would be needed in the future.

The Ford-Ferguson tractor with the three-point hitch was also eating into sales. International Harvester design engineers had to come up with something fast if they were going to

stay competitive. The long-standing litigation between Ford and Ferguson had finally been settled and International Harvester was able to breathe a sigh of relief and take a hard look at their own hitch development. It's difficult to believe, but International Harvester designers seemed to have missed an important market signal.

The I-H answer to the Ford-Ferguson system was a two-point hitch they called the "Fast Hitch." While their hydraulically operated Fast Hitch system was capable of managing some implements, it was incompatible with all of the three-point implements that were now becoming the industry standard. International Harvester would finally be forced into line by industry groups, such as the American Society of Agricultural Engineers, who were defining the industry's standards. By 1958, International Harvester finally offered a three-point hitch.

Some observers say that it was about this time that International Harvester began to lose control of the marketplace. After dominating the market for decades, I-H appeared to be unable to meet the technology challenges. There were few significant changes to the I-H product line after 1947 and even fewer after the Ford-Ferguson lawsuit was settled.

Some critics say that I-H did little more than change the model designations in 1954, hoping the marketplace would think that International Harvester was matching the competition. So, like

John Deere's, the I-H product line was renamed, changing the old alpha designations to numbers. In 1954, the International Harvester models A, C, H, and M were replaced by models 100, 200, 300, and 400.

There had been one bright spot in 1947. The Farmall Cub was introduced, a cute little machine that stood knee-high to a grasshopper. It was designed to compete with the little Deere L, but strangely enough, John Deere discontinued their L in 1947. The Farmall Cub looks a great deal like the Model A, and they are sometimes difficult to tell apart.

The Cub was designed for nurseries and folks with large yards rather than farm acreage. Rated under ten horsepower, it could pull a single plow. Seven or eight implements were custom built for the little Cub, and the machine could be equipped with a hydraulic Touch Control system. Because of its size and the selection of implements, the Cub is a favorite of collectors. Besides, they are so small, collectors can fit three or four Cubs in their garage.

But other than the Farmall Cub, there was little excitement from International Harvester. They kept a close eye on John Deere, matching their marketing strategy. When John Deere went to number designations for their models in 1952, I-H went to numbers in 1954. When John Deere went to a two-tone green and yellow paint job in 1955, I-H changed their paint scheme in 1956 to a two-color red and white. International Harvester was losing market share, and John Deere would take over the lead as number one in tractor sales.

International Harvester would continue to suffer, as an economic turndown and then a labor union strike by the United Auto Workers took large chunks out their corporate hide. Finally, in 1989, they were purchased by Tenneco, a conglomerate that already owned the J.I. Case Company. This extraordinary American company is now partnered with another agricultural institution. They are known today as Case I-H.

# Selecting the Finest—
# The Collectors

Some folks collect stamps, some collect paintings. All over our agricultural landscape there is a new type of collector, someone who finds and restores farm tractors. The stamp collector and the art accumulator seldom have to face the challenges encountered by the average tractor enthusiast. Many of the really interesting and valuable machines need thousands of hours of careful restoration before they can even be shown privately, let alone exhibited.

Tractors tend to be ignored by most urbanites—city folks seem to focus only on collecting and restoring automobiles. But it was the farm tractor that helped make urbanization possible by producing enough food to feed the city. An American invention, the traction engine, now known as the farm tractor, changed world agriculture and established America's position in the global economy.

The tractor made every farmer an entrepreneur rather than a serf. It was the tool that provided the basis for the American economy, a system that encouraged independence. And it was the tractor that promoted free enterprise and rewarded individuals who were willing to risk becoming working capitalists.

The tractor complemented the Homestead Act, finally making it really possible for a single farmer to manage 160 acres (64ha), sometimes more. The tractor was a truly American invention, one developed through the talents of dozens of immigrant blacksmiths, machinists, and craftsmen. Once developed and refined, this machine was immediately exported—thousands of traction engines were shipped to Europe and South America even before World War I.

It sometimes seems that members of the current generation discount any invention from before their own lifetime, thinking their grandparents lived in the Dark Ages of pretechnology. They ignore—or forget—that early American machinists and engineers were an extremely inventive group of individuals.

Today, thousands of farm tractors are being rediscovered, restored, and collected. They are cherished for their history, for their beauty, and for the personal challenges they present to the restorer—it takes a great deal of patience and know-how to restore a hunk of junk that has weathered thirty winters in an Iowa hedgerow. But determination and persistence are more important assets to a tractor collector than skill.

What determines a "collectible" tractor? Tractor collectors use most of the same guidelines that other antique collectors follow. To qualify as a bonafide antique, a tractor should be at least forty years old. Early machines and those with a low serial number for the particular model are especially prized. Certain types of tractor technology and mechanics—the early International Harvester "friction drive," for example—are especially intriguing and collectible.

Like any antique, rare examples can be expensive. A Hart-Parr "Little Devil" went for $33,000 at an auction in the summer of 1996, but a rare Hart-Parr 20-40 costs nearly twice that. Also especially valued are machines in very good condition, with the original paint. A glorious Persian orange Allis-Chalmers can be a big winner. But so can a very rare model of the John Deere HWH.

Real tractor enthusiasts, however, have always been more than just collectors. A tractor is made to run, to be a useful and productive tool—it's not just another pretty face. Many collectors restore their machines to use—some for everyday chores, others for tractor pulls and competitions, still others for demonstrations and exhibitions.

Finding a rusty scrap heap in a ditch with a tree growing through the frame, a collector lovingly disassembles the remains of the carcass and hauls it back to his shop, garage, or kitchen table. Then, bit by bit, he puts the tractor back together. Sometimes it takes thousands of hours over several years. But it is always a labor of love.

Rescuing a tractor is an opportunity for every farm boy to become a real Prince Charming. After fighting through the undergrowth of brambles in the hedgerows, shrugging off the naysayers and scoffers, painstakingly tracking spare parts from all sections of the countryside, the collector spends hours upon hours assembling a workable machine. Finally, the long-awaited moment.

Carmin Adams collects and restores small tractors like this Allis. Because he specializes in smaller models, his complete collection of compact machines is comfortably stored at his home.

Bending over the sleeping beauty, he carefully sloshes some gasoline in the carburetors and gently turns the flywheel. With a flutter and a cough, the tractor awakens to his loving touch. And man and machine live happily ever after.

## Iron Men and Frail Machines

There is a simple maxim about tractor collectors: they are as diverse as the models they collect; and most have a favorite model or models. Some live in town, some live in the country. They are old or young, male or female, wealthy or struggling. Some collectors have a national reputation and hundreds of machines. Most will tell you the story of their adventure discovering each tractor, where it was found, and how it was restored. Sometimes they will even tell you how much they paid for each tractor.

## Carmin Adams—Small Tractors

Tractors, tractors, everywhere—in the garage, in the carport, and under a tarp next to the house. While many tractor collectors have their machines housed in barns or sheds on their rural property, Carmin doesn't have this luxury of space: he lives in town. His house is on a steep hillside overlooking the Santa Clara Valley, once the finest agricultural area in California. Although there are fruit trees nearby, Carmin has no room to field-test his collection.

Carmin Adams builds fine machinery as a hobby, and his shop is neater than most people's kitchens. Retired from a lifetime of work as a service engineer for automobiles, Carmin really knows machinery. When Carmin retired he looked forward to doing restoration, but living in town limits the size of the tractors he can keep, so he decided to collect and restore small tractors.

Carmin began his collection with two Farmall Cubs, small tractors that first appeared just after World War II, designed to compete with the little Fordson and the John Deere models L and LA. When Carmin finished restoring the two Cubs, he looked for another interesting small tractor to fix up. He started working on an Allis-Chalmers Model B, and soon he had an opportunity to help an acquaintance work on an Allis-Chalmers Model G. Carmin says that the Model G "was such a beautiful little tractor that I could hardly wait to start working with it. And I could hardly leave it alone once I did start on it."

The Model B is a four-cylinder machine, sold to farmers between 1937 and 1957, and designed to be a powerful little machine for the small farm. In contrast, the Model G was designed for the truck farmer or nursery owner, someone with a small acreage of high-value crops. The Model G came out in 1948 and it is a rare machine, looking like a miniature road grader without a scraping blade.

After Carmin finished working on the two Farmall Cubs, a John Deere LA restoration candidate arrived. The John Deere Model LA was also designed for the farmers with little land: those who grew specialty crops or cultivated less than an acre. Introduced in 1941, Models L and LA are sometime called the John Deere "Little." When Carmin started working on his Model LA, the work went quickly. It wasn't in bad shape, he remembers, especially when compared to the two Farmall Cubs. He notes that always he tries to clean up as he works and makes an effort to maintain a shop that is environmentally correct.

Carmin Adams is running out of room in the garage, but he has his heart set on finding a Cub Cadet for his collection. He will be careful about selecting his restoration candidate, and is willing to pay a little more for a machine in better condition. His outstanding collection is a good example of what can be accomplished without a large shop, an overhead hoist, and a grease pit. With careful planning, it is possible to collect and restore a notable variety of tractors, even in the heart of town.

# JR Gyger—Case

JR Gyger, a retired farmer from Lebanon, Indiana, currently owns the premier collection of Case orchard tractors. JR has been acquiring tractors for years, but he never considered himself a "collector" until very recently. He observes that when farmers own a good, working machine, they try to hold onto it, if there is space in the barn.

Known as JR, the nickname given to him by several older sisters, he says he has been interested in tractors all his life—Gyger can't recall the first tractor show he ever attended, it was so long ago. These days his tractor collection and his retirement time are split in two. His "summer" collection of Case tractors remains on the farm in Lebanon just northwest of Indianapolis; his "winter" collection, the Case orchard tractors, reside in Florida.

JR bought his first Case tractor when he graduated from high school: a brand new 1952 Case DC. He bought it to farm with and found it to be a very solid machine. His farm business prospered, starting with a modest forty acres (16ha) and growing to include 1,200 acres (480ha) of corn, beans, and wheat. The machines that made the Gyger farm such a successful operation were Case tractors. JR recalls that when he first started farming, it was the dealer who kept him buying Case machinery. "He always treated me right," said JR, "and I stayed with him until he retired. Case tractors were pretty popular in this part of the country because this dealer put a lot of them out."

JR did try a John Deere tractor before he retired from farming. He wanted to see if the Green Machine was as great as some people were saying, but he decided that while the two machines were about equal in performance, the Deere tractor was hard to get into. "A John Deere is okay for somebody who just don't know any different," is his final judgment.

Since JR always farmed with Case machinery, he already had many machines from his own career to use as a foundation for his collection. He now has about seventy-five Case tractors altogether, fifty of which are completely restored. He

jokingly comments that there are always a few in need of restoration, staring him in the face. JR's goal is to have a complete set of Case orchard machines, and he is just about there.

Although he hadn't used them on his own farm, after JR retired he became interested in orchard tractors and actively searched the citrus groves in Florida for old machines. Orchard tractors were only useful in orchards and vineyards, since they were designed to protect fragile tree limbs from injury during cultivation, so few orchard models were offered by any major tractor manufacturer. This scarcity is a major factor in their appeal to collectors.

Case orchard tractors were the first to use inflated rubber tires. An experimenter in the early 1930s installed modified aircraft tires on a tractor for better traction and handling in the sandy Florida soils, and this innovation led to the development and adoption of pneumatic tires just for

Some tractor enthusiasts focus their collection on one particular type of tractor. JR Gyger specializes in J.I. Case orchard tractors, and he owns nearly a complete set of these beautiful tractors with their big flaring wheel covers.

tractors. The bright colors of these orchard tractors seem to complement the citrus groves where they were used, but Flambeau Red and Desert Sand were standard colors on all Case models of the era.

JR sometimes compares the people who collect Case tractors to other tractor collectors. "Case collectors," he says, "have a lot more to talk about than just brass tags and four bolt pedestals. Deere collectors just don't have too much to talk about. The two-cylinder motor was the same for years. When John Deere collectors put their machines out on display, all they have is a bunch of green two-cylinder tractors."

# Doug Peltzer—John Deere

Pioneers of California citrus farming, the Peltzer family has been growing oranges for three generations. The family first farmed in Orange County, California, but moved to Porterville in the foothills of Tulare County in the 1940s. Doug Peltzer says he has always like tractors, but he didn't realize he was a collector until long after they moved to Porterville. One day the mailman mistakenly delivered a copy of a magazine for tractor collectors. Suddenly Doug realized he was not alone; there were many other folks out there who loved tractors. He joined the local chapter of EDGE (Early Day Gas Engine) Association and began collecting seriously.

Today Peltzer is noted for his outstanding collection of John Deere tractors. But the first machine he bought and restored as a card-carrying collector was an early Fordson. He recalls that the Fordson was considered a very advanced machine for its type, but Doug knew it was not the best tractor around. In fact, it had some serious design deficiencies. The Fordson had no governor on the engine and the driver's seat was located just over the transmission, giving the operator a very hot ride.

Doug had already acquired a number of old tractors, machines he bought because he was intrigued with their technology. After he restored the Fordson he evaluated his collection and decided that he needed a John Deere. He soon found an unstyled 1936 John Deere Model A and a companion 1935 Model B. It was the acquisition of the two John Deere tractors that kindled his passion for Green Machines.

Doug admits that at first he was not too selective about collecting Deere tractors. He bought any John Deere he could find, and many people who knew he was interested just gave him their old machines. Within three short years he owned nearly sixty John Deere tractors, mostly Model As and Bs with a few exotic models in the herd. He noted that this was just before the explosion of collector interest in the machines.

Peltzer feels fortunate to have acquired most of his more unusual machines between 1987 and

1991, when machines were still readily available. His first goal as a John Deere collector was to find one machine of each model in the letter series. Having quickly accomplished that goal, he decided to add John Deere machines that represented a significant milestone in technology. Today his John Deere collection is focused primarily on one-of-a-kind and representative models of John Deere tractors.

Although Peltzer owns some rare John Deere machinery, his collection also includes other important early tractors. He acquired all of his machines because he was intrigued by their mechanics and the technology they represent. His 1918 Heider has a friction drive and moves either forward or in reverse by engaging one side or the other of the flywheel. His 1920 Best crawler represents a milestone in technology that remained virtually unchanged until 1951. His patiently assembled collection includes a 25-50 Avery, a 28-50 Hart-Parr, and a Rock Island 12-20. There is also a very special McCormick-Deering 10-20, a machine with sentimental ties since it belonged to Doug's grandfather. And every John Deere collector needs a Waterloo Boy: Doug owns a 1920 model in very good condition.

## Looking Ahead: What's Over the Horizon for the Tractor Enthusiast

What does the future hold for the tractor industry and for tractor collectors? These are two very different questions, and each one requires a separate answer.

We all like to eat, every day if possible. It's as simple as that. Since tractors are tremendously efficient farm tools, they are still needed to produce our daily bread. Older tractors, including many of the machines built thirty or forty years ago, are still going strong all over the world. Simple to operate, burning a variety of fuels, and easy to repair, many old American tractors are still doing a full day's work. New tractors will continue to appear on farms in the wealthier

nations, helping to increase agricultural productivity. So it looks like tractors, new and old, will be around for many decades to come.

The appearance of serious and ambitious tractor collectors, however, is a relatively recent phenomenon. Although tractor shows and competitions have been around since the first steam-powered machines chugged across the Great Plains a century ago, the organized collecting of tractors is a relatively recent occurrence.

Today's collector has an interest in machinery and a fair amount of mechanical expertise, in order to evaluate machines for any repairs, as well as the free time and flexible schedule needed to enjoy the hobby. The requisite ingredients seem to be a little leisure time, some mechanical ability, reasonably good health, a bit of extra cash, and an enormous amount of affection for old iron. The current generation of tractor enthusiasts are retiring earlier and living longer, a combination that gives them a formidable edge in assembling their collections.

The future of tractor collecting seems limitless, as dozens of converts discover the satisfactions of restoring and operating old machines. It is a complex hobby that appeals to the mechanic as well as the historian. It is a hobby as suited to the collector with a small space and a limited budget as to the individual with a thousand acres and deep pockets. Pound for pound, some of the rare little collectible model-tractor toys are more valuable than the real ones that are still operating in the back forty. So as long as there are tractors, there will undoubtedly be collectors. And the wonderful advantage of owning most of these interesting machines is that—when times get really tough—the tough can take to the field on their favorite tractors!

# Bibliography

Broehl, Wayne. *John Deere's Company.* New York: Doubleday & Company, 1984.

Buescher, Walter M. *The Plow Peddler.* Macomb, Ill.: Glenbridge Publishing Company, 1992.

Collier, Peter, and David Horowitz. *Fords: An American Legend.* New York: Simon & Schuster, 1987.

Currie, Barton W. *The Tractor.* Philadelphia, Pa.: The Curtis Publishing Company, 1916.

Erb, Dave, and Eldon Brumbaugh. *Full Steam Ahead: J.I. Case Tractors and Equipment 1842–1955.* St. Joseph, Mich.: American Society of Agricultural Engineers, 1993.

Gray, R.B. *The Agricultural Tractor— 1855–1950.* St. Joseph, Mich.: American Society of Agricultural Engineers, 1975.

Holmes, Michael S. *J.I. Case: The First 150 Years.* Racine, Wisc.: The Case Corporation, 1992.

King, Alan C. *Massey-Harris: Data Book No. 6.* Delaware, Ohio: Massey-Harris Company, 1992.

Leffingwell, Randy. *Caterpillar.* Osceola, Wisc.: Motorbooks International, 1994.

McMillan, Don. *John Deere Tractors and Equipment: Volume One: 1837–1959.* St. Joseph, Mich.: American Society of Agricultural Engineers, 1988.

Wendel, C.H. *Encyclopedia of American Farm Tractors.* Osceola, Wisc.: Crestline Publishing Company, 1979.

———. *Nebraska Tractor Tests Since 1920.* Osceola, Wisc.: Crestline Publishing Company, 1985.

Wik, Reynold M. *Benjamin Holt and Caterpillar.* St. Joseph, Mich.: American Society of Agricultural Engineers, 1984.

Williams, Robert C. *Fordson, Farmall and Poppin' Johnny: A History of the Farm Tractor and Its Impact on America.* Urbana and Chicago, Ill.: University of Illinois Press, 1987.

If you would like more information about tractors, their history, and their collectors, you might consult one of the following magazines devoted to the subject.

*Antique Power Magazine*
Pat Ertel, Editor
P.O. Box 838
Yellow Springs, OH 45387

*Green Magazine* (John Deere)
R. and C. Hain, Editors
Rural Route 1
Bee, NE 68314

*M-M Corresponder* (Minneapolis-Moline)
Roger Mohr, Editor
Route 1, Box 153
Vail, IA 51465

*9N-2N-8N Newsletter* (Ford)
G.W. Rinaldi, Editor
154 Blackwood Lane
Stamford, CT 06903

*Old Abe's News* (Case)
David T. Erb, Editor
Route 2, Box 2427
Vinton, OH 45686

*Old Allis News* (Allis-Chalmers)
Nan Jones, Editor
10925 Love Road
Belleview, MI 49021

*Red Power* (International Harvester)
Daryl Miller, Editor
Box 277
Battle Creek, IA 51006

# Photo Credits

*Antique Power Magazine:* pp. 16, 17, 29, 102

©Hans Halberstadt: back endpaper, pp. 1, 2, 7, 11,12–13, 13 right, 14, 15, 18, 19, 22, 23, 24, 26, 27, 28, 30, 32-33, 33 right, 34, 36, 37, 38, 43, 44–45, 45 right, 46, 47, 48, 50, 52, 53, 54–55, 55 right, 57, 58, 59, 60, 61, 66, 68, 69, 70, 72, 73, 74-75, 75 right, 76, 77, 78–79, 80 both, 81, 82, 83, 84, 86–87, 87 right, 89, 90, 91, 92, 93, 94, 96, 97, 98, 99, 100, 101, 103, 104, 105, 106, 107, 108–109, 109 right, 111, 113, 114

Courtesy J.I. Case: front endpaper, pp. 9, 10, 20, 35, 39, 40, 41, 49, 64, 65, 67, 85

Courtesy John Deere:  p. 62–63

Courtesy Richard Walker:  p. 21

# Index